Strung Together

Strung Together

THE *Cultural Currency* OF *String Theory* AS A *Scientific Imaginary*

Sean Miller

The University of Michigan Press • *Ann Arbor*

Published in the United States of America by
The University of Michigan Press
Manufactured in the United States of America
♾ Printed on acid-free paper

2016 2015 2014 2013 4 3 2 1

A CIP catalog record for this book is available from the British Library.

Library of Congress Cataloging-in-Publication Data

Miller, Sean, 1969–
Strung together : the cultural currency of string theory as a scientific imaginary / Sean Miller.
pages cm
Includes bibliographical references and index.
ISBN 978-0-472-11866-3 (hardback : acid-free paper) — ISBN 978-0-472-02896-2 (e-book)
1. Science—Literary collections. 2. String models. 3. Science in popular culture. 4. Science—Language. 5. Physics—Philosophy.
I. Title.
PS509.S3M55 2013
810.8'0356—dc23

2012042608

Acknowledgments

This book is the result of five years of sustained effort. Much of this work took place in an environment of quasi-monastic solitude—hour upon hour ensconced in a carrel at various libraries in London and New York, in a cublicle at Paragraph NY, or in a nicely cooled office at Nanyang Technological University in Singapore. There I would peck away on an old netbook whose hard drive clacked like the spindly legs of what I sometimes imagined to be a mechanical spider. Scholarship can be lonely, even if its content is, in the abstract, part of a noble conversation that leaps across borders, oceans, and time itself. Progress would not have been possible without the support of many people—to whom I am extremely grateful.

First and foremost, I want to thank my wife, Shveta. When I began this book, we were dating. That our subsequent engagement and marriage coincides with the steady progress made on this project is, to me, no coincidence. Your loving presence in my life is a continual source of motivation and inspiration for me, vocational and otherwise.

I also want to express gratitude to Jeff Miller and Susan Sealy, whose support on many levels means a great deal to me. My heartfelt appreciation also extends to Shveta's family, to Hayley and James Harrigan, whose hospitality in London has been exceptionally helpful, to my Birkbeck comrades, Michael Cade-Stewart and Matt Wraith, and to Jerry Mundis, whose advice on how to establish sound writing habits has been instrumental.

A hearty thanks is owed especially to Andrew Gibson for his expert guidance and wise counsel. I also want to thank Adam Roberts and the rest of the English Department at Royal Holloway. Many thanks go out to the NTU Division of English, where I have been serving as a postdoctoral fellow in contemporary literature, and, in particular, to Neil Murphy, Terence Dawson, and Daniel Jernigan. I also would like to thank two members of the NTU Division of Physics, Chew Lock Yue and Rainer Dumke, whose suggestions on improving the manuscript have been immensely helpful. I very much appreciate the valuable contributions to this project that the following people have made: Laura Salisbury, Daniel Cordle, Donald Petcher, Bruce Clarke, Linda Henderson, Ronald Schleifer, and Priscilla Wald.

Last, I want to thank the staff at the University of Michigan Press, and especially Tom Dwyer, Christopher Dreyer, Alexa Ducsay, Christina Milton, Michael Kehoe, and Shaun Manning.

Contents

CHAPTER 1

Introduction

String Theory as a Theory of Everything

String theory is reputed to have begun in 1968, when a postdoctoral fellow named Gabrielle Veneziano, working at CERN,[1] one of the world's leading high energy physics laboratories, proposed a solution to a vexing problem concerning the interaction of subatomic particles in the nuclei of atoms. He accomplished this by using a formula he had found in an eighteenth-century mathematics text.[2] Two years later, three other theorists—Yoichiro Nambu, Leonard Susskind, and Holger Nielsen—independently suggested that Veneziano's redeployment of this antique mathematical function implied that the particles that formed the nuclei of atoms were not actually zero-dimensional point-particles, but rather, extended out into an extra dimension.[3] While Nambu and Nielsen described this hypothetical object as a "harmonic oscillator"—a term common to both classical mechanics and quantum theory, Susskind was the first to liken it to a vibrating string ("Dual" 483). As many popular accounts go, in what one might describe as a leap of imagination, a radical reconceptualization of the ordinary string was adapted to theoretical high energy physics in order to replace the point-particle of quantum theory and serve as the fundamental constituent of the universe.

In essence, string theory declares that the cosmos is made up of strings: they are either open or closed, possess tension, and vibrate. The degree of tension corresponds to the frequency of vibration which, in turn, determines what form a string takes. These strings are miniscule—on the scale of 10^{-33} centimeters.[4] When experimental physicists attempt to observe them with their instruments, the colliders and detectors at CERN and other such laboratories—that currently can probe distances of about 10^{-17} centimeters—they appear, so the theory contends, as the myriad point-particles so accurately delineated by quantum theory. As of yet, strings have not been observed in nature.

Since 1968, string theory has blossomed into a fully fledged attempt to reconcile the two established theories of matter and force in the universe—quantum theory and general relativity—into one mathematically consistent and empirically valid formalism. Public awareness of string theory grew tremendously with the publication of the first book-length popularizations in the late 1980s. Thereafter, string theory popularizations have been published—and enthusiastically consumed—with increasing regularity. One can explain this in part due to several factors: the radical transformation of "common-sense" that string theory as a theory of physical reality suggests; the controversial nature of string theory's status as science; and perhaps most significantly, the tradition within contemporary Anglo-American culture for theoretical physicists to serve as the custodians and purveyors of a particularly resonant form of fundamental truth; namely, that which constitutes the cosmic order.[5] Having captured the popular imagination, one now observes string theory imagery cropping up with increasing frequency in both popular culture and literature.

This book represents the first investigation into string theory as a specifically cultural phenomenon. String theory, in particular, provides a fruitful topic for this sort of approach for two important reasons. Firstly, it is a relatively new theory, having only been studied for about forty years. String theory has begun to seep into public awareness only in the past twenty-five years or so, beginning with the publication of Michio Kaku's *Beyond Einstein* in 1987, followed by F. David Peat's *Superstrings and the Search for the Theory of Everything* and *Superstrings: A Theory of Everything?*, edited by Paul Davies and Julian Brown, both published in 1988. String theory's public profile grew immensely in 1994, with the success of Michio Kaku's follow-up, *Hyperspace*. As the subject of cultural study, string theory represents, to borrow a favorite trope of the popularizers themselves, relatively unexplored territory.

Secondly, string theory would seem, by its nature, to lend itself to literary adaptation: as a theory of everything, it is currently the leading scientific attempt to unify into one consistent theoretical framework the known matter in the universe with the known forces. As such, it would potentially form the foundation for all other scientific knowledge.[6] In this sense, theoretical physics—which deals, in all its mathematical and imaginative esotericism, with the world on the largest and smallest of scales—has always been fertile ground for a cultural imagination hoping to bind the cosmos to human-scale, quotidian concerns. Given string theory's even further remove, it would seem to be all the more vulnerable to such imaginative appropriation.

This book's core approach is to examine the cultural currency of string theory by means of a concept I call a *scientific imaginary*.[7] Drawing primarily on the work of French philosophers of science Michèle Le Doeuff, Gaston Bachelard, Michel Serres, and the American cognitive linguist George Lakoff, the following chapter defines this pivotal term *scientific imaginary*, in part by contrasting it to a scientific realism that sorts "pure" concept from "playful" or "illustrative" image. In a scientific enterprise such as string theory, an imaginary is a complex of images that, while ostensibly set off from concept, nevertheless problematizes it. These images serve as doubles for mathematical arguments, yet they also possess their own inferential relations distinct from this structure of correspondences. In this sense, a congeries of images that in one valence are bound individually in one-to-one correspondence to mathematical expressions, in another valence, together constitute an autonomous whole, an imaginary. Such an imaginary is marked as scientific inasmuch as it retains its reference to what a culture recognizes as scientific practice—in the case of string theory, with a particular, dynamically defined amalgam of methods and reasons, empiricisms and logics, with an emphasis on the plural. The following chapter offers several examples of scientific imaginaries in an epistemological and cosmological context in order to demonstrate how the concept can be employed to analyse textual presentations of string theory. This chapter shows that, in effect, the work of a scientific imaginary is fourfold: one, it grounds scientific concept in familiar human-scale experience; two, it mediates between human agency and the agency of the objects a theory manifests; three, it invests those objects with substance; and importantly, four, it situates scientific theory within a larger imaginative context. Without this connecting tissue, so to speak, scientific knowledge such as string theory would have no epistemological footing.

In addition to its timeliness, part of this book's value lies in the fact that it brings to bear on string theory, and importantly, its technical discourse, a novel method of close reading that synthesizes developments in contemporary cognitive linguistics with more familiar literary critical techniques. Furthermore, it offers new insights into the complications that arise in the consumption by a non-specialist audience of string theory popularizations—insights that pose important implications for the consumption of science popularizations in general. Ultimately, this book suggests a novel way for scholars, critics, and lay readers to contextualize and evaluate the adaptation of scientific ideas by popular culture and literary texts.

This book makes four key claims. First, that the imagination is central to the production of string theory as scientific knowledge. Second, we must read string theory popularizations carefully, since the imaginary that they present often engenders a particular ideological bias. Third, there are clumsy and nuanced ways for imaginative writing to adapt string theory ideas, and by implication, ideas from theoretical physics in general. And fourth, that string theory as an imaginary reflects, paradoxically, both an archaic sensibility and a contemporary culture where the ethos of globalization dominates.

As mentioned earlier, since technical articles serve as the main vehicle for the production of string theory as scientific knowledge, a principal concern of this investigation will be, in chapter 3, a careful consideration of them. I have chosen these articles because they represent key episodes during string theory's historical development. Significantly, I will focus on the expository prose that surrounds and frames the core mathematical content of these key technical articles. Such a reading strategy readily concedes that the ostensibly conceptual content of the technical articles; namely, the mathematics, speaks to an exclusive and highly specialized audience. The authority to judge the legitimacy of the science as it is practiced through mathematical argument within string theory technical articles rests with the professionals themselves. Nevertheless, the chapter examines several crucial instances where images—specifically the string, brane, and extra dimension—enter the technical discourse within a specific article, and thus contribute to and, in many respects, serve to determine, an imaginary. Chapter 3 also investigates the extent to which such an imaginary can be said to cohere from article to article—through the technical discourse as it develops historically. Since images refer to specific mathematical expressions, a key question will be: to what extent can the complex of these images, an imaginary, be said to cohere? Furthermore, how do the terms and the images associated with these terms come to gain a common cultural currency? How do they become conventionalized as a specifically epistemological imaginary, as a means of knowing the world?

With popularizations—the subject of chapter 4—the dynamics of expression change. In my survey of string theory popularizations, I have found them to fall into three basic categories: one, accounts written by string theorists that take a positive, relatively uncritical stance; two, accounts written by physicists who do not specialize in string theory but who treat it as a secondary topic, with varying degrees of critical distance; and three, accounts that are openly critical of string theory.[8] Chapter 4 examines the first type of popularizations for the following reason: they are heavily invested in string theory's scientific legitimacy—and in its ultimate institutionalization—and therefore are highly interested in communicating string theory as a coherent, meaningful theory. Yet, since they omit the mathematics out of explicit consideration for an audience they assume is mathematics-averse, an imaginary takes precedence over pure conceptual content.[9] In these popularizations, the exposition of string theory comes to be, in many respects, a vehicle for the substantiation of what Michel Serres calls an "endo-epistemology" (*Conversations* 128). A way of knowing the world becomes a self-contained, all-encompassing means of structuring the relationship between an imagined human agency and a radically remote and alien cosmos.

Strictly speaking, from a realist perspective, a discourse such as that constituted by string theory popularizations would be suspect if it made claims about the objective world without direct recourse to an agreed-upon empiricism and a self-consistent logical structure—the purview of specialists. The rest of us—a lay readership, including the scientists who specialize in other fields—are simply not in a position to confirm the legitimacy of string theory's empirical or mathematical consistency. In effect, the non-specialist reader must take the truth-value of the content of a popularization on a certain form of faith endemic to contemporary Anglo-American culture; that is, faith in the authority of "hard" science to present the heretofore "hidden" reality of our world. Accordingly, to accept string theory as a legitimate science is, to a certain extent, an act of belief. Chapter 4 investigates the extent to which the imaginaries in three string theory popularizations, which are representative of the first type, are, as Caroline Jones and Peter Galison put it, "a revealed truth otherwise hidden," dependent wholly on the authority of the narrator, or conversely, something "self-contained because it 'speaks for itself'" (354).

The 1999 publication of Brian Greene's *The Elegant Universe* represents another watershed moment in popular awareness of string theory. When organizing a collection of creative writing inspired by string theory, entitled *Riffing on Strings* (2008), I noticed that many authors credit Greene specifically with introducing them to it.[10] In *Science on Stage*, Kirsten Shepherd-Barr draws attention to the recent surge in output and reception of science plays. A noteworthy subgrouping of the trend of the past decade in science plays are those that are specifically string theory-themed. Chapter 5 represents the first sustained attempt at critical attention to imaginative texts, including some of these plays, which engage with string theory. For the purposes of this investigation, I will define imaginative texts as texts where the imagination is understood to be primary, as with drama, fiction, and poetry. The following are issues relevant here: the salient differences and similarities between the imaginary such texts present in relation to popularizations and technical discourse, and the extent to which these texts, in effect, reproduce both a string theory imaginary and an endo-epistemology that supports and sustains that imaginary.

Together, this constellation of texts—the manifold technical articles, popularizations, and imaginative treatments—come to constitute a string theory imaginary. While this imaginary is uniquely heterogeneous—a miscellany of images, internal relations amongst those images, and correspondences between the images and an objective world, supposedly manifest through mathematics, beyond its own structure—within this imaginary, a pattern of flows emerges. Images move unidirectionally from technical exposition to popularizations to the literary. The concluding chapter explores the implications of understanding a form of scientific knowledge such as string theory as an imaginary in a wider sociological context. The discussion relates string theory as a late twentieth and early twenty-first-century social phenomenon to broader cultural trends; for example, globalization. In tracking the migration of images from one discourse to another, I will pay particular attention to the role an imaginary plays in the transmission, substantiation, and adaptation of a science such as string theory to contemporary Anglo-American culture, considered broadly. We will then be in a position to ask: to what extent does a string theory imaginary constitute another supporting "pillar" in the edifice of scientism, where scientism is understood as an unreflective faith in the monologic omnipotence of science? Or does this string theory imaginary—with all its ambiguities and ambivalences—serve to frustrate scientism?

By ordering the chapters as I have (an introduction to string theory, a chapter on the concept of a scientific imaginary, string theory technical discourse, popularizations, then imaginative writing), I do not mean to imply the prescriptive reinforcement of an established hierarchy of scientific knowledge. Rather, as a scholarly monograph, this book is descriptive insomuch as it respects the mandate to provide credible evidence in accounting for the intertextual flow of string theory ideas and images from one discourse to another. Accordingly, it attempts to account for the actual flow of string theory images and ideas as it has occurred over the course of the past four decades, which has been, as previously stated, predominantly unidirectional.

As chapter 5 shows, much contemporary imaginative writing that engages with string theory tends to reproduce this conventional hierarchy, in large part, by deferring reflexively to string theorists' cultural authority as guardians of cosmic truths. Yet I do not see imaginative writing as doomed categorically to being only a weak reflection of string theory technical discourse. Through a close reading, in particular, of Carole Buggé's play *Strings* and Brenda Hillman's poem "String Theory Sutra," chapter 5 explores how imaginative writing can and, in these noteworthy cases, does challenge the epistemological assumptions of scientific realism, while nevertheless avoiding the lazy distortions of some representations of string theory. Lastly, the concluding chapter provides an account of how string theory itself may be shaped imaginatively by the historical and cultural technoscientific milieu out of which it has emerged.

One intriguing possibility would be to rearrange the contents of the book so that its structure embodies its claim about the primacy of imaginaries in string theory as a scientific culture. Such an approach would reflect a sensibility that emphasizes the importance of play in its methodological framework as an implicit epistemology, in keeping with one of the book's important influences, Michel Serres. This would be an altogether different book, one more akin to a philosophical essay in the style of Serres. Yet I suspect that a hypothetical Serres-esque essay that endeavored to foreground non-scientific cultural influences on string theory as the technical discourse of a professional community (for example, how science fiction may have shaped the work of the prominent string theorist Ed Witten) would have a difficult time justifying its evidence and, as a consequence, would risk making the kinds of catachreses that the likes of Alan Sokal find so farcical. As such, the ordering of the book reflects, in some respects, a methodical pragmatism more in keeping with the style of another of its key theoretical sources, George Lakoff.

String Theory's Emergence

For those readers who may be unfamiliar with string theory, what follows in this introductory chapter is an account of the key features of string theory within the context of its historical development—in effect, how string theory attempts to reconcile and thus move beyond its predecessors—general relativity and quantum theory. To avoid complication, this description of string theory hews closely to the standard narrative found in most string theory popularizations. While both general relativity and quantum theory have achieved canonical status within the high energy physics community, string theory remains very much a work in progress. As such, this chapter also touches upon the ongoing debate in the broader theoretical physics community concerning string theory's legitimacy as science.

Thomas Kuhn writes of science progressing not holomorphically, as a smooth and continuous process of accretion, but rather in a series of fits and starts, of one "paradigm" of "normal science" being ousted by another in a paroxysm of revolutions or "shifts" (6, 10). String theory might exemplify just such a paradigmatic shift—upending one "conceptual box," as Kuhn puts it, one set of theoretical commitments, in favor of another (5). Broadly speaking, string theory concerns itself with the interactions that occur between the fundamental objects of nature, described in terms of fundamental forces. Established physical theory currently posits four fundamental forces: electromagnetism; the strong nuclear force, which acts to bind the particles that comprise the nuclei of atoms together; the weak nuclear force, which contributes to radioactive decay; and gravity.

As we shall see in chapter 4, string theorists themselves often narrate the history of the theory as a romance where the heroic physicist struggles to reconcile all the known particles with all the known forces—a noble quest for that ultimate prize, a "theory of everything." It is important to acknowledge, though, that when string theorists speak of a theory of everything, they mean something specific to the problems that the profession considers legitimate: their notion of "everything" does not imply, at least directly, every domain of scientific knowledge (e.g., applied physics, astronomy, chemistry, biology, and so on), but rather the problem of conceiving a quantum theory of gravity. A quantum theory of gravity would integrate the formalism and the body of experimental evidence that constitutes quantum theory, the currently accepted theoretical explanation for, loosely speaking, the realm of the very small—subatomic matter, which is subject to electromagnetism, the weak and the strong nuclear forces—with the theory of general relativity, which currently is employed most fruitfully in treating gravity in the context of interactions of large objects: apples, planets, stars, galaxies, the entire universe. This is a specific concern with specific problems relating to attempts at synthesizing the mathematical formalisms of quantum theory and general relativity. The attempt to integrate the "conceptual box" of quantum theory with the "box" of general relativity must yield a precise formalism. This formalism would represent a third, entirely distinct "conceptual box," with its own cache of legitimate problems, its own set of concisions and symmetries. This third paradigm, though, must be backward-compatible, to borrow an expression from software engineering: it must not contradict any of the demonstrated accuracies of its predecessors with respect to observed phenomena. Yet historically, this requirement of backward-compatibility becomes ever more stringent as solid evidence in diverse experimental domains mounts. As such, the problem of fitting a new paradigm to the growing and seemingly incompatible bodies of evidence requires ever more sophisticated formalisms. String theorists today wrestle with the logical structure of their core conceptual commitments, its fundamental objects and how to reconcile these objects with the requirements of quantum theory, to retrofitting the framework of quantum theory onto the string, so to speak, while simultaneously coaxing the string into the formal constraints of the tensor algebra and Riemannian geometry of general relativity.[11]

With that in mind, let us backtrack briefly in order to describe the historical and conceptual context out of which string theory has emerged. Since string theory concerns itself with unifying the known fundamental forces of nature, an appropriate starting point would be the mid-1800s. While Sir Isaac Newton had formalized a theory of universal gravitation in 1687, it was James Clerk Maxwell, building on the work of Michael Faraday, who proposed in 1865 a theory to account for what were then recognized to be two other forces of nature distinct from gravity—electricity and magnetism. Experiments demonstrated that electricity and magnetism behave in ways that the Newtonian theory could not explain adequately, limited as it was to describing the world in terms of rigid bodies and the "invisible" force of gravity that instantaneously attracted them. Furthermore, Faraday observed that electricity and magnetism both behaved in a remarkably similar fashion. Maxwell was able to model the behavior of electricity and magnetism through a set of equations (known, appropriately, as *Maxwell's equations*) that mathematically expressed the concept, not of an instantaneous force interanimating rigid bodies, but of a pervasive, disembodied *field*.[12] Maxwell's was a theory of electromagnetism, in which this force permeates absolutely flat and infinite Newtonian space as a harmonically oscillating field with associated values for frequency, momentum, and charge—positive, negative, or neutral. Maxwell went on to make the prediction that there are electromagnetic waves that travel at the speed of light and that also possess light's properties of polarization. Heinrich Hertz confirmed his prediction experimentally in 1888; light has since been considered a particular manifestation of electromagnetic energy—with distinctly field-like behavior.

In 1905, building on the work of Maxwell, Hendrik Lorentz, Henri Poincaré, and Hermann Minkowski, among others, Albert Einstein addressed an important problem that Maxwell's hypothesis concerning light had opened up in Newtonian mechanics. In the Newtonian universe, momentum is infinitely cumulative: in theory, if a person travelling near the speed of light flicks on a flashlight, the light that emerges from the bulb should travel at the speed of light plus the speed at which the person is moving. But since light was and is known to possess a finite, constant speed, any scenario with a speed beyond the speed of light is physically impossible. Einstein proposed that the accumulation of speed in the physical universe does not behave as Newtonian theory predicts: it behaves asymptotically—speed always approaches a maximal limit, the constant speed of light in a vacuum, measured at approximately 186,000 miles per second. Or, as Kaku concisely puts it, "The speed of light is the same in all constantly moving frames" (*Hyperspace* 82). Among other things, Einstein's solution—his theory of special relativity—expresses this limit on the possibilities of speed. Einstein realized that this has startling implications for our understanding of time. In *The Elegant Universe*, Greene offers a vivid illustration of one of these counterintuitive implications using two astronauts in space, named George and Gracie, who are equipped with space suits and special jet packs.

> [I]magine that the relative speed of George and Gracie when they pass and are moving apart is 99.5 percent of light speed. Further, let's say that George waits 3 years, according to his clock, before firing up his jet-pack for a momentary blast that sends him closing in on Gracie at the same speed that they were previously moving apart, 99.5 percent of light speed. When he reaches Gracie, 6 years will have elapsed on his clock since it will take him 3 years to catch her. However, the mathematics of special relativity shows that *60* years will have elapsed on her clock . . . In a real sense, George's motion has made him a time traveler, albeit in a very precise sense: He has traveled into Gracie's future. (44–45, emphasis in original)

According to special relativity, Greene summarizes, since "observers in relative motion will have different perceptions of distance and time . . . time slows down when an object moves relative to us because this diverts some of its motion through time into motion through space" (25, 50). Both time and space are, in effect, integrally pliant phenomena: space-time. We do not notice this relationship in our daily experience, because the motion we perceive never approaches the speed of light.

In 1907, Minkowski formulated special relativity in terms of a four-dimensional geometric, which has since come to be known as *Minkowski space-time*. What distinguishes Minkowski space-time from its predecessor—the Euclidean space used by Newton—is the affording of time a space-like status. In Newtonian mechanics, time remains isolated as a phenomenon. In Minkowski space-time, time forms an integral part of the total "fabric" of space. In this sense, Minkowski space-time is not simply a physical space, but an epistemological abstraction: a manifold that captures the counterintuitive interrelationship between space and time. Special relativity presents the image of the universe as a dynamic, interrelated whole, often imagined as a fabric.[13] According to the theory, the essence of this cosmos, as a system, is a particular kind of change; namely, the force of movement as rotation not just through space, but time as well. Since no moving body can travel faster than the speed of light (for if it were to reach the speed of light, time would stop), light itself forms an absolute boundary in Minkowski space-time. This is often called in the theory a "light cone," which defines the limits of causation, since an object may never contact or communicate with another object outside this cone.

In 1915, Einstein expanded the space-time of special relativity to incorporate acceleration.[14] With his theory of general relativity, he posited a fundamental equivalence between acceleration and the force of gravity within a space-time framework. A famous equation, the Einstein Field Equation, expresses this relationship using tensor algebra. A tensor describes a collection of quantitative attributes that are associated with particular coordinates in the geometric space of the system. The Einstein Field Equation equates energy and momentum (the stress-energy tensor) with space-time curvature (known generally as the Einstein tensor). Unlike special relativity—with its flat Minkowski space-time—the geometry of general relativity is Riemannian: the tensor of information having to do with stress-energy, with the density and flux of energy within the system, manifests as the curvature of space-time itself, not just the motion of objects through it. Space-time bends and warps in dynamic response to the motions and densities of energy and matter that constitute it. General relativity has helped make numerous predictions that have been borne out in subsequent experiments; for example, that the universe is expanding; that it originated with a Big Bang; and that, given enough matter concentrated in one region in space, gravity there can become so powerful that not even light can escape, forming a black hole.

In the 1920s and 1930s, while Einstein's theories of special and general relativity were forcing a radical revision of Newtonian cosmology, an international cadre of physicists, led by Einstein, Max Planck, Ernest Rutherford, Niels Bohr, Erwin Schrödinger, Werner Heisenberg, Paul Dirac, Max Born, and Louis de Broglie, among others, was focusing its attention on theorizing the realm of the very small. Out of their collaborations emerged quantum theory, which offered up its own paradigm shifts to problematize Newtonian theory. If we include the contributions of the physicists that succeeded that initial generation, including Richard Feynman, David Bohm, Murray Gell-Mann, Sheldon Glashow, Steven Weinberg, Abdus Salam, and many others, we can summarize that quantum theory[15] has made the following contributions.

(1) The concept of the *quantum*, or the discretization of energy. In quantum theory's historical development, this eventuated in the disarticulation of the atom into a seemingly ever-expanding array of subatomic particles—with exotic names like muon, tau, neutrino, and charm quark—organized within the standard model into groups based on symmetries, or shared attributes.[16] The fundamental image of the quantum particle is a point in space with no internal extent. In addition to having a variable position and momentum, particles may possess other distinguishing attributes—constants such as rest mass, spin (angular momentum), charge (positive, negative, or neutral), color (red, green, or blue), and flavor (up, down, strange, bottom, top, or charm).[17]

(2) Along with electromagnetism, the discovery and incorporation of two other forces: the weak nuclear force, which contributes to a certain form of radiation decay, and the strong nuclear force, which binds the nuclei of atoms together. It is important to note that current versions of quantum theory do not account for the force of gravity. So, in a significant sense, it is not comprehensive.

(3) Wave/particle duality, which asserts that subatomic quanta behave both like waves and like particles. The wave behavior of the quanta is expressed in the theory in terms of *wavefunctions*, which describe the probability distributions that define a particle and that provide predictions for the given attributes, such as position and momentum, about to be observed.[18]

(4) Quantum theory further formalizes this characteristic wave/particle ambiguity through the Heisenberg uncertainty principle, which attests that, because of a given particle's dual nature as both wavefunction and point-particle, there exists an inversely proportionate possibility of measuring with precision either the particle's position or its momentum. The Heisenberg uncertainty principle, generally applied, can also pertain to the relationship between a given particle's energy and time, as well as other specific pairs of attributes. In effect, it demonstrates the fundamental inability to know with absolute precision a pair of attributes of a given particle, if those attributes have a mutually dependent relationship within the mathematical formalism of the theory.

(5) Consistent with the ambiguity implied by wave/particle duality, quantum theory can describe forces in terms of particles as well; these force particles are sometimes called "messenger" particles, treating the force act-

ing between two "matter" particles as if it were a "packet" of information. For example, the theory can describe the strong nuclear force that binds quarks within the nucleus of an atom as the exchange of a "messenger" particle called a *gluon*.

(6) Quantum theory's earliest incarnation, quantum mechanics, assumes a background of absolute space and a distinct and absolute time dimension; it does not incorporate the constraints of special or general relativity. Contemporary quantum field theories do incorporate special relativity, thus a relativistic, albeit flat, four-dimensional Minkowski space-time. But, as mentioned earlier, no widely acknowledged quantum theory accounts for gravity—neither the force itself, hypothesized as a messenger particle called a *graviton*, nor the effects of the curvature of space-time in general relativity.

In addition to being an exceedingly effective explanation for the dynamics of the subatomic world, quantum theory has contributed or given outright birth to many important technologies including applied chemistry, nuclear fission, the laser, the diode, the electron microscope, and the transistor, integral to all electronics—including the computer. Yet in spite of quantum theory's demonstrated predictive accuracy, many theorists would agree that quantum theory feels somewhat like a kluge,[19] with its unwieldy proliferation of subatomic particles organized into symmetry groups,[20] the seventeen or so numerical constants that must be inserted into the formalism and that the theory has no way of explaining through its own internal logic,[21] the qualification that only quantum theories that are "renormalizable" yield workable results,[22] and lastly, the unsettling epistemological ambiguity, previously mentioned, between wave and point-particle states. Such theoretical deficiencies leave most theorists (and a fair share of experimentalists as well) unconvinced that, with quantum theory to explain the subatomic realm and general relativity to describe the universe as a whole, we currently have in our possession a true theory of everything.

The universally accepted standard model of quantum theory came to maturity in the 1970s. Since then, much theoretical inquiry has concerned itself with the margins of the two canonical theories; in effect, with scenarios that arise when general relativity and quantum theory come into direct conflict. These scenarios occur when physical phenomena are simultaneously very large (e.g., high energy densities) and very small. Two such scenarios that often occupy contemporary high energy physicists are black holes, where massive amounts of matter get compressed into a quantum-scale space, and the split second immediately following the Big Bang, where the entire universe was compacted into a comparably microscopic extent. In these physical extremes, both theories break down and, as such, provide a strong incentive for theorists to concoct a new theory—one that reconciles quantum theory with general relativity and thus would allow them to explore with greater predictive power these exotic physical scenarios—black holes and the very early universe.

String Theory's Development

The string theory proposed by Nambu, Susskind, and Nielsen in the early 1970s attempts to explain the mysterious dynamics of the strong nuclear force. Their theory is called *hadronic string theory*. The protons and neutrons that make up the nuclei of atoms are classified as *hadrons*, those subatomic particles that experience the strong nuclear force. Yet in the early 1970s, the most promising theory for explaining the strong nuclear force was quantum chromodynamics, a point-particle field theory. Quantum chromodynamics posits the existence of a new category of particles called *quarks*.[23] Quarks possess a novel form of charge, arbitrarily designated "color"—hence the "chromo-" in chromodynamics. Through a battery of new experiments that took advantage of more powerful accelerators, the high energy physics community came to the consensus that quantum chromodynamics was overwhelmingly more successful at explaining the strong nuclear force than hadronic string theory.

Nevertheless, a small group of theorists persisted in their expansion of string theory. In 1974, John Schwarz of the California Institute of Technology and Joël Scherk of the Ecole Normale Supérieure proposed that string theory was not merely a theory of the strong nuclear force, nor even of the more comprehensive grouping of subatomic particles that comprise the standard model. In their reformulation, known as *bosonic string theory*,[24] they argued that one of the string's vibrational patterns possesses properties that correspond to the then posited messenger particle for gravity, the graviton. Since bosonic string theory incorporates the force of gravity through the graviton, Schwarz and Scherk felt they could claim that theirs was a legitimate quantum theory of gravity; it predicted gravity.

Subsequent investigation showed, however, that their revision and expansion of string theory suffered from its own slew of inconsistencies. For example, in order to be consistent with quantum theory, bosonic string theory requires not the four space-time dimensions of relativity, but twenty-six space-time dimensions.[25] Needless to say, for most physicists committed to the principles of scientific realism—however philosophically nuanced—

this was a difficult pill to swallow. Scientific realism holds as a fundamental tenet that there is an objective world independent of our capacity to know it. Should overwhelming evidence gathered using a diversity of methods confirm the existence of a phenomenon in the world, it ought to be taken to be objectively real; for example, that the universe consists of three spatial dimensions: length, height, and breadth. In addition, bosonic string theory predicts the existence of a new kind of particle, called a *tachyon*, that possesses the reality-defying feature of negative mass—or, from a complementary perspective, propagation beyond the speed of light.[26]

Meanwhile, experiments with particle accelerators in the mid-1970s were probing the internal structure of hadrons and finding them decidedly point-like at the scale of about 10^{-16} centimeters. This was glaringly inconsistent with the proposed size of the string in bosonic string theory, which, at 10^{-16} centimeters, should reveal its extended dimensionality. Throughout the rest of the decade, experimenters confirmed, time and again, the predictive accuracy of the standard model of quantum theory. Accordingly, quantum theory became further institutionalized, while string theory languished in relative obscurity on the peripheries of accepted practice.

In string theory lore, this all changed in 1984 when Michael Green at Cambridge University, who had been working with John Schwarz at Caltech for several years on improving the theory, published what Brian Greene has dubbed a "landmark" paper.[27] The paper claimed to resolve the problems that had marginalized string theory for the past ten years. Many string theorists refer to the appearance on the scene of this revised theory by Green and Schwarz as "the first superstring revolution"—the moment when string theory became a legitimate contender for the highly anticipated "theory of everything." Green and Schwarz accomplished this by incorporating supersymmetry into the theory (hence the "super" in "superstring"). Supersymmetry is a principle that provides for the pairing of fundamental particles, organized in the standard model into the two moieties of bosons and fermions,[28] in a mathematically coherent way such that many of the inconsistencies in quantum field theory—or string theory—may cancel out. The standard model accounts for its spectrum of subatomic particles by sorting them into the three distinct symmetry groups that are juxtaposed. Supersymmetry merges these three groups into one "supergroup." Supersymmetry also suggests the possibility of the unification of all four fundamental forces into what is sometimes called *supergravity*.[29]

As a consequence of the manner in which it organizes the particles into one "supergroup," supersymmetry predicts the existence of a new array of partners for the known quantum particles—what are sometimes referred to as *sparticles*.[30] The incorporation of supersymmetry into Green and Schwarz's superstring theory requires the model to reduce the number of posited space-time dimensions from twenty-six down to ten. It also does not generate particles of negative mass, the so-called tachyons. Another hallmark of their new superstring theory is that it recalibrates the string tension from the relatively large (and untenable) scale of the original bosonic string theory down to about the Planck scale, or 10^{-33} centimeters. Green and Schwarz's particular version of superstring theory, which has since come to be known as Type I superstring theory, also predicts the graviton, thus reinvigorating string theory's claim to be a theory of quantum gravity.

Shortly after Green and Schwarz's theory reignited a burst of interest within the high energy physics community, other theorists proposed alternative and potentially competing versions of superstring theory—whose composition depends largely on how supersymmetry is incorporated into the theory's overall structure. These competing versions of superstring theory are known, respectively, as Type IIA, Type IIB, Heterotic SO(32), and Heterotic $E_8 \times E_8$.[31] Further work in the late 1980s and early 1990s showed that, although they share certain features, the various theories diverge enough to suggest a significantly contrasting picture of physical reality. Brian Greene describes the mood within the string theory community since then: "This has been an embarrassment for string theorists because although it's impressive to have a serious proposal for the final unified theory, having five proposals takes significant wind from the sails of each" (*Elegant* 284). For physicists, one criterion by which they judge a theory is that it has a certain inevitability; that the theory reproduces precisely the specificity of observed phenomena—in the case of string theory, that it generates the exact particle spectrum of the standard model, or a justifiably extended version thereof. That five plausible versions of string theory have come to coexist suggests that these theories, although they share a strong conceptual continuity, may also have built into their internal logic too many degrees of freedom, too many possible permutations and thus physical consequences—a feature that directly contradicts the principle of inevitability. The embarrassment for string theorists lies in these versions' excessive flexibility. They lack the requisite rigidity for generating (or retroactively predicting) the observed particularities of the physical world. Speaking of general relativity, Lawrence Krauss writes: "The complexity of the theory means that we still have not yet fully understood all its consequences; therefore we cannot rule out various exotic possibilities" (*Physics* 34). This observation could apply to string theory as well. The multiplicity of—as yet, not disproved—versions, coupled with an overall lack of understanding as to their physical implications, contributes to a climate within string theory where "exotic possibilities" cannot be ruled out.

Work on the five competing versions of string theory continued more or less independently until 1995, when, at the premier annual string theory conference (called *Strings* and held that year at the University of Southern California), Edward Witten, from the Institute for Advanced Study, presented a paper[32] that launched what the community has come to regard as the "second string theory revolution." Witten's solution to string theory's embarrassment of riches is to suggest that its five versions are actually different perspectives, symptoms of the perturbative methods employed in the theories' mathematical formalism, that all point toward one unified framework, which he dubs "M-theory."[33] He proposes that the ten space-time dimensions called for in the earlier theories are also an approximation; M-theory requires eleven.[34] With this and other adjustments, Witten claims that the five theories (and one more called eleven-dimensional supergravity) can be organized into one meta-theory, due to relations that depend on certain mathematical constants and the geometry of space-time itself. As one of its entailments, Witten's revisioning of string theory posits a central role for drumskin-like objects that extend into two or more spatial dimensions—what are called *membranes*, or simply, *branes*.[35] Branes behave similarly to strings. Some more recent versions of string theory even define the string as a *one-brane* and a point-particle as a *zero-brane*.

It is important to note here, though, that Witten himself, along with the rest of the string theory community, is acutely aware of M-theory's limitations. M-theory gestures toward a quantum theory of gravity, but does not actually codify it. Much of the work in the field since 1995 has focused on the search for a more coherent and precise formulation of M-theory. M-theory has also spawned its own revisions. For example, in 1996, Cumrun Vafa, of Harvard University, published a paper that proposed an "F-theory," which calls for twelve rather than eleven space-time dimensions, along with other modifications in the geometric structure of its space-time.[36] As we shall see later, when one considers the arguments of string theory's detractors, the efforts of the past fifteen years, though resulting in a veritable deluge of papers, have yet to provide a third "revolution"—one that can claim to offer a formalism that possesses the required inevitability for a universally convincing model of physical reality.[37]

While some string theorists work on refining M-theory, another subdiscipline attempts to reconcile the currently conceived formalism of the theory with known astrophysical phenomena. A paper authored by Juan

Maldacena of the Institute for Advanced Study epitomizes such efforts.[38] Maldacena argues that if the geometry of a certain ten-dimensional string theory model is reconfigured, it becomes consistent with the holographic principle, which stems from studies of information loss in black holes and posits that a physical model of $(n - 1)$ dimensions can correspond exactly to a physical model of n dimensions; an effect analogous to a holographic projection. For the purposes of the discussion at hand, what is significant here is that string theorists are actively borrowing concepts from astrophysics in order to make their models more robust, so that they better corroborate more experimentally grounded physics. Another example of this is a paper[39] published in 1996 by Vafa and Andrew Strominger, also of Harvard University, that undertakes a similar calculation of the information states of a certain type of black hole using a string theory formalism.

These types of approaches seek to mold string theory such that it better conforms to astrophysics. Other approaches attempt to inject string theory into astrophysical models. Shortly after Witten's revolutionary M-theory proposal of 1995, Joseph Polchinski, of the University of California, Santa Barbara, posited the existence of another structure where open strings are fixed onto a time-like membrane, which he called a "D-brane."[40] An alternative cosmological model based on Polchinski's concept of the D-brane has complicated conventional Big Bang theory. In a D-brane-based scenario, the four space-time dimensions of our universe may exist as a "braneworld" within a larger and enveloping eleven-dimension metaverse, often called *the bulk*. One version of braneworld cosmology has replaced the Big Bang with a "Big Splat"; two braneworlds cyclically collide to provoke the expansion of our braneworld universe.[41] Other speculations occupy themselves with concepts such as tears in the "fabric" of space-time, wormholes, infinitely extended extra dimensions rather than miniscule, curled-up ones, or gravity "leakage."[42]

Yet another subdiscipline of string theory that has emerged post-M-theory aims to integrate the theory more fully with general relativity, beyond predicting the existence of the graviton (the force of gravity) as the consequence of one particular string-vibrational mode. Like special relativity, M-theory's precursors assume an essentially flat space-time background.[43] As discussed earlier, general relativity shows that space-time exhibits curvature. The very topology of space-time warps and ripples in direct relation to its own gravitational pull; that is, in proportion to its energy density. Yet a quantum theory of gravity implies a fully quantized universe, where space-time itself is no longer infinitely divisible and infinitely extensive, but rather, granular and finite. The underlying gravitational field, composed of a multitude of gravitons linked together in a kind of matrix or fabric, constitutes space and time. Were we able to probe this fabric at a sufficiently tiny scale, we would encounter lacunae in space itself—a thoroughly counterintuitive notion.[44] Work concentrating on such a granular space-time fabric has led string theorists to attempt to constitute space-time itself with a certain form of zero-brane,[45] whose behavior at the Planck scale, the theorists contend, is best described not by the commutations and anticommutations of Riemannian geometry, but by a formalism developed in the early 1990s by the French mathematician Alain Connes called *noncommutative* geometry.[46] Edward Witten has also attempted to integrate M-theory with another background-independent space-time theory based on general relativity—Roger Penrose's "twistor" theory.[47]

String Theory's Status as Science

While string theory has grown, in some sense, to dominate the discipline of theoretical physics, it has also managed to attract an increasing number of critics. Almost unanimously, they point out that string theory seems to concern itself only tangentially with observational data, and as such is more an imaginative speculation than an empirically based science. String theorists would not disagree that this lack of testability is a direct consequence of the scale of the fundamental string—the Planck scale. Current accelerator technologies are only capable of probing scales about a thousand times smaller than the atomic nucleus. The Planck scale is smaller than that by a factor of more than a million billion—analogous to inspecting with a telescope individual atoms in the Andromeda galaxy, which is approximately 2.5 million light years from Earth (Krauss, *Hiding* 209). Even the Large Hadron Collider at CERN is unable to probe such remote scales.[48] Some experimentalists calculate that in order for an accelerator/collider to have the capacity to probe the Planck scale, it would require more energy than all the energy in the entire known universe combined.[49] Others contend that the energy required merely outstrips several times over the total power output available on Earth, assuming current technologies and resources.

Theorists such as Brian Greene hope that the Large Hadron Collider will serve the purpose of further legitimizing string theory, either by confirming the existence of sparticles, the particle duals predicted by supersymmetry, or by providing solid evidence for the existence of extra dimensions.

Nevertheless, although string theory attempts to incorporate supersymmetry, it is not the only theory to do so. Many high energy theorists would consider the existence of sparticles merely circumstantial evidence.

String theorists also suggest that astrophysics might provide evidence for the existence of strings. Some astrophysicists currently study the cosmic background microwave radiation that persists as a kind of residue or record of the explosive expansion of the universe immediately following the Big Bang. Certain interpretations of string theory suggest that this explosive early universe may have left traces of superstrings stretched out to macroscopic proportions—cosmic strings. New instruments such as LIGO (Laser Interferometer Gravitational-Wave Observatory) and LISA (Laser Interferometer Space Antenna), designed to detect gravitational waves, may help to confirm the existence of such cosmic strings.[50] One group of physicists has proposed an experiment to probe, albeit indirectly, extra dimensions larger than the Planck scale, yet still microscopic—a scenario that certain versions of string theory predict.[51]

Several prominent physicists have become outspoken critics of string theory, including Roger Penrose, Sheldon Glashow, Lawrence Krauss, Philip Anderson, Lee Smolin, and Peter Woit.[52] Many detractors evoke the Popperian argument that string theory, since it is so far removed from experimental observation, has the inexcusable quality of being unfalsifiable.[53] They also point to string theory's embarrassment of riches—the nearly endless proliferation of allowed physical scenarios due to the theory's excessive flexibility, its myriad degrees of freedom. To claim legitimacy on the grounds that certain string theory versions can generate "semi-realistic" particle spectrums strikes these critics as entirely unconvincing.

Some theorists evoke the "anthropic principle" in their defense of string theory's seemingly excessive degrees of freedom.[54] One prominent braneworld scenario suggests that the "bulk" metaverse may contain upwards of 10^{500} four-dimensional braneworlds, such as our universe, each with their fundamental physical laws tuned slightly differently—this metaverse being termed "the Landscape." Given such a metaverse, proponents argue that we find ourselves in a universe with the precise configuration of physical laws that can support the formation of galaxies, stars, planets, life, and ultimately, sentient life, simply because we are here to observe just that. Since we exist, such a universe, tuned precisely to accommodate our existence, must be possible. The anthropic principle is what Susskind would call an "environmental" argument; it suggests no inevitability based on first principles as to the particular physical properties of our observable universe. These properties are merely a product of circumstance.[55] Many physicists, including some string theorists, find such anthropic arguments to be entirely unsatisfactory. They expect a theory to express a more stringent degree of inevitability, to explain fundamental causal relationships in a self-consistent manner.[56]

String theory also poses many unresolved and seemingly intractable mathematical impasses. What follows is a brief summary of these issues. No one has been able to formulate a string theory using a non-perturbative method,[57] which most view as essential to making the theory coherent on extremely small scales. Furthermore, M-theory itself has yet to be successfully quantized—a necessary prerequisite for full validity. The theorists specializing in string *field* theory, a version of string theory that attempts to integrate background independence,[58] similarly, have made little tangible progress. Furthermore, string theory has yet to establish the fundamental inevitability of the string itself. Strings, whether open or closed, along with branes of various dimensions and configurations, emerge from various theories depending on how they are formulated. No version of string theory demonstrates a self-evidently particular composition of such fundamental objects. Furthermore, a fundamental principle such as the equivalence principle[59] in Einstein's general relativity has yet to be formulated within the context of string theory. Ed Witten has suggested that string theory might confirm the notion that space-time itself is an emergent phenomenon, not a fundamental assumption.[60] Rather than "plugging" a space-time background into the formalism as an initial condition, string theory ought to be able to produce space-time through its own machinations, and thus, be able to "predict" the existence of space-time. Krauss points out that string theory might make a genuine prediction with respect to one epistemological puzzle still very much unresolved—that of calculating the energy of empty space itself, or what is often called in the literature *dark energy*.[61] But string theory has yet to do so.

Finally, there is the problem mentioned previously as to the physical status of string theory's proposed extra dimensions. Barton Zwiebach writes:

> In superstring theory a similar calculation fixes the dimensionality of spacetime to the value of $D = 10$. The fact that string theory cannot be a good . . . quantum theory in any arbitrary dimension shows that string theory is very constrained. Even more, since the dimension of spacetime is uniquely selected by the requirement of consistency, we can say that string theory predicts the dimension of spacetime! (231)

While bosonic string theory requires twenty-six space-time dimensions, superstring theory calls for ten—the $D = 10$ of the calculation to which he refers. What Zwiebach finds so remarkable is that by forcing superstring theory to conform to the requirements of quantum theory and special relativity, in order to be mathematically self-consistent, its space-time background must have ten dimensions. In effect, the mathematics of superstring theory emphatically predicts that the cosmos must possess these extra dimensions.

Critics such as Penrose and Krauss argue that these extra dimensions may be nothing more than "mathematical artifacts." Krauss asks:

> What is the utility of an extra hidden dimension if ultimately nothing is hidden except the existence of the extra dimension? And what is the practical meaning of extra dimensions if you can experience all there is to experience without actually moving into them?[62]

Krauss seems to suggest that string theorists, by arguing for the physical existence of extra dimensions, have fallen into that epistemological trap, described by Gaston Bachelard, where "what is real but hidden has more content than what is given and obvious" (*New* 31–32). In effect, form takes precedence over substance in the string theorists' implicit epistemology. The theoretical formalisms have more reality than what we can interact with physically, directly. It is sufficient to "grasp" theoretically the string in order to grant it the status of being real.

Yet our bodies move in three dimensions: up and down, left and right, back and forth. As such, an emphatically shared and intuitively self-evident experience compels us to acknowledge that we live in a physical universe of three spatial dimensions. String theory conceptualizes time in a comparable manner to special relativity. Within the formalism, time is defined as a space-like dimension—an axis, much like the forward and back our bodies locomote along, composed of an array of points. These points are measured out into a metric relative to the motion of an arbitrarily chosen body (for the metric of seconds, etc., the motion of the earth around the sun). Within the theory, the idea of a spatial dimension, a fixed line of sight that allows one to locate objects in space (and time) becomes readily multiplied beyond the limits of commonsense physical space. Counterintuitively, we must extrapolate extra dimensions from the familiar back and forth, up and down, and left and right motion of bodies through space. In a sense, it is impossible to imagine a fifth, a tenth, or twenty-sixth dimension in their own "literal" terms. Our imaginations, grounded as they are in our bodily experience, are simply not equipped to process such an alien abstract space.

The mathematics of extra dimensions itself is not such a novelty. Mathematical operators, such as the Hamiltonians employed by quantum theory, frequently make use of them to map extra-spatial physical attributes such as spin and charge to spatial dimensions. Special relativity formulates the physical world in extra dimensions—four total, three of space, and one of a space-like time. As noted previously, this counterintuitive binding of space with time constitutes, in and of itself, an abstract space. Yet theoretical physicists in general would be loath to concede that the extra dimensions of special relativity also represent what Krauss calls "mathematical artifacts." In order to offer any legitimate promise of expanding our capacity to intervene in the causal structure of the physical world, the formalism of string theory must necessarily account for a massive amount of empirical data, gathered from a vast body of preceding theory and experiment. Most of this observational data is only indirectly accessible to our apprehension, by way of mediating instruments. Even these instruments, accelerators and colliders such as the Large Hadron Collider, yield abstract data that must be reconciled to the formal framework. Yet it would seem that a string theorist's insistence on the existence of these extra dimensions results from an allegiance to a particular epistemology that I have called *realist.* The three chapters that follow explore certain significant features of this realist epistemological commitment on the part of string theorists and the consequences that such a commitment has on the status of the theory's fundamental objects as objective phenomena—strings, branes, and extra dimensions.

To echo Krauss, even if some battery of future experiments—such as the search for sparticles or macroscopic extra dimensions—were to confirm string theory's prediction that the universe has ten (or eleven, twelve, or twenty-six) space-time dimensions, that would not necessarily invalidate a debate concerning the epistemological status of these extra dimensions and what implications they would pose for our understanding of string theory's claim to objective reality. Positive results from such experiments would only confirm string theory's efficacy—that it provides accurate prompts for intervening in the causal structure of the world, prompts for an intervention still ultimately grounded in embodied experience. And we live with the practical certitude that our bodies exist in three dimensions: they move up, down, left, right, back and forth. To perhaps state the obvious, a dimension is an abstraction that formalizes this motion in such a way that we may project the event of bodily motion into remote perceptual terrain. The potency of this abstraction, originating from the concept of a line of sight or the ordered rows and columns of an agricultural field, lies in its capacity to accommodate the additional information needed to describe accurately the full range of attributes of the phenomena within these remote perceptual domains.

Physical theories would seem to function admirably in the following fashion: a theory reconceptualizes physical reality in a precise way; physicists then draw inferences and design experiments based on that reconceptualization—what Kuhn calls the work of "normal science." The experiments, satisfactorily diverse in methodology and independently repeated, then either validate or falsify the theory. What we could only imagine and never experience directly attains the status of stable scientific knowledge. Accordingly, string theorists claim that the precise formalism of the theory predicts the existence of extra dimensions. The history of theoretical physics in general has demonstrated repeatedly the efficacy of this approach. The broad acceptance of a revolutionary theory is predicated inevitably on its making a bold prediction that contradicts the prevailing intuition—the accuracy of which is subsequently confirmed through some newly fashioned experiments.

The initial institutionalization of general relativity is a prime example. Sir Arthur Eddington's 1919 expedition to the African island of Príncipe to observe a solar eclipse confirmed Einstein's prediction concerning the shifting of light due to the gravitational influence of massive bodies such as the sun. Thereafter, theorists and experimentalists labored to design further experiments that took as their template the warped and pliant spacetime formalized by general relativity. String theorists practice a specific form of epistemology—with its own covenant of behavioral and symbolic conventions, validation procedures, and legitimate problems—that privileges a categorically objective reality, or to use a conventional image, a "deeper truth." While consensus is relatively easy to build with respect to the self-consistency—or correctness—of a given piece of mathematics, as critics complain, string theorists currently struggle to produce more than just flashy mathematics, however self-consistent. They have yet to provide experimentalists with more emphatic direction on how to corroborate the theory with evidence. Yet in spite of its lack of a champion such as Sir Arthur Eddington, who legitimized the then seemingly implausible theory of general relativity, string theory enjoys a great deal of popular recognition, if not implicit support. In lieu of evidence, string theorists themselves, along with a public that ultimately funds their livelihoods through institutions such as universities, would seem happy to see their investigations continue on the basis of two things: one, a body of interesting mathematical problems to solve; and two, as the following chapters detail, a compelling imaginary to describe the cosmos.

Those keen to philosophize quantum theory have focused principally on its most counterintuitive idea—the dual nature of its fundamental wave/particle as elucidated, in part, by the Heisenberg uncertainty principle. Meanwhile, once enough experimental evidence had accumulated, the correctness of quantum theory's formalism, for instance, the Schrödinger wave equation, was never much in doubt. To this day, the high energy physics community may use quantum theory without being able readily and coherently to imagine it.

In its current state, string theory's continuing institutional prestige is complicated by the dubious empirical status of the science itself. Those critics anxious to fix string theory's meaning to a political agenda have dismissed it as "pomo" junk science—the product of a self-involved cabal of mathematics-intoxicated speculators. More moderate detractors argue that it lacks a sufficient degree of inevitability to justify all the resources and manpower devoted to it. They point out that, at best, string theory is "semi-realistic" in its backward-compatibility—its capacity to replicate the canonical standard model. Although string theory excites many with its promise of incorporating, along with supersymmetry, the force of gravity into quantum theory, one critical inference that follows from the string's formal structure; namely, the incredibly minute scale of strings themselves, makes any potential experimental validation difficult to anticipate.

String theorists argue that the existing body of experimental evidence and the formal constraints of the mathematics they employ serve to discipline their speculative impulses, to bring them back down to earth. They insist that string theory's formal component dictates that it must progress by a consensus beholden to the strictest of rules. Yet, as the following chapters explore in more detail, the formalism's imaginative complement may very well serve as a repository for an *in*formal content for which the mathematical arguments themselves cannot account. String theory's formal complexity, its exclusionary technical opaqueness, its facility for speculative model building, and its tenuous status as science all conspire to leave it ripe for a reading that calls attention to the pivotal role the imagination plays in its technical exposition and, following on from that, in its wider cultural currency.

As we have seen, to evoke string theory as a unitary whole obscures its bewildering heterogeneity. One might speak of a plurality of string theories, save for the fact that certain more recent versions no longer even posit the string as their fundamental object. When cataloguing string theory, one is obliged to include M-theories, F-theories, brane theories, and Landscape theories, among others. As U.S. Supreme Court Justice Potter

Stewart famously wrote of pornography in *Jacobellis v. Ohio*, in lieu of a precise definition—"I know it when I see it."[63] The use of the single term *string theory*, which stands for a complex of concepts and ideas, would seem to be merely a matter of linguistic convention and practical convenience. Nevertheless, in light of this problem of multiplicity, the close readings that follow of string theory-themed texts will explore the extent to which continuity in the presentation of string theory as an imaginary does indeed suggest a uniform string theory imaginary as a cosmic order.

CHAPTER 2

A Return to the Eleventh Dimension

String Theory as a Scientific Imaginary

Now If You Believe in Physics, You Got the Eleventh Dimension . . .

[Scott] Mehring, of Mechanicsburg, Pennsylvania, is forty-eight years old, the onetime owner of a business that had something to do with performance cars. He wore a tight leather motorcycle jacket with no visible shirt underneath and had a Rod Stewart haircut. He liked to party, he told me, and was ready to go out and party hard, but because he'd lost his license for various reasons he had no car and his cab had not yet arrived. So, sure, he'd be happy to share his views with me. I took out my recorder.

"If you go back to the Big Bang," he said, speaking rapidly, "the elements, I'm not sure exactly what they actually were, but whatever the elements were—the atom, the neutron, the proton neutron, whatever it was that created the Big Bang—*where* did that stuff come from? Spontaneous generation is a dead theory—at one time they thought it was true—left a piece of meat on the ground maggots appeared, they thought the maggots came out of the meat, but actually they came out to eat the food, so you can't say spontaneous generation created it . . . Now if you believe in physics, you got the eleventh dimension—it's a new theory, the eleventh dimension—and inside the eleventh dimension they say that there's an infinite number of universes. So my take is that if you die on the earth, we just somehow hop over to the eleventh dimension, and hop from universe to universe to universe forever inside the eleventh dimension. So that means the Bible could be right with everlasting life after we die. *But*, okay, the elements that started the Big Bang, if that was an intelligent designer? Then you've got another complication. If there was, like, one dude somewhere at the very top that

> created everything? Well, *where did he come from*? Who created him? And who created the God who created God? It gives me goose bumps. It's a loop, like in computer programming—it's an endless loop."
>
> He paused and shook his head. His cab had arrived.
>
> "If you think about this too much," he concluded, "you can go insane." (Chapman 63, emphasis in original)

The passage is an excerpt from an article by Matthew Chapman in *Harper's Magazine* entitled, "God or Gorilla: A Darwin Descendant at the Dover Monkey Trial." On first inspection, the interviewee, Scott Mehring of Mechanicsburg, Pennsylvania, would seem to be indulging in a facile form of philosophizing. In a *tour de force* of pastiche, through his evocation of the eleventh dimension, Mehring touches with amusingly egregious inaccuracy on a diverse array of scientific disciplines, including cosmology (the Big Bang), quantum theory (protons, neutrons), biology (the medieval theory of spontaneous generation, a textbook anecdote almost universally foisted upon American children in grammar school; and implicitly, evolution), theology (the Bible, eschatology, and intelligent design), computer science (feedback loops), psychology (the perils of insanity), and most notably for this book's purposes, string theory. As described in the previous chapter, "the eleventh dimension" as a scientific concept originates in M-theory—a version of string theory formulated in 1995 by Ed Witten in an effort to integrate the five distinct and competing versions of superstring theory. M-theory suggests that our universe might consist not of the familiar three dimensions of space and one of time, but of eleven dimensions—ten of space and one of time. The various explanations for the existence of these extra spatial dimensions depend on string theory's mathematical formalism. But in the absence of such formalisms, "the eleventh dimension," as Mehring calls it in his labyrinthine conjecture on life, the universe, and everything, offers the imagination a fertile and mutable space for projection.

Mehring's speech is framed by an ironic distance. Chapman is reporting on the *Kitzmiller v. Dover Area School District* trial that took place in Pennsylvania in 2005, where "eleven parents sued to remove intelligent design from the curriculum" (54). Chapman, the great-great-grandson of Charles Darwin himself, met Mehring outside of his hotel on the last day of the trial. Mehring was not directly involved with the trial, but for Chapman, he presumably represented a telling example of that distinctly American curiosity—the "average Joe." Without explicitly renouncing journalistic impartiality, coded here through a mention of the recorder, Chapman conditions the reader's initial impression of Mehring through a careful selection of details: the tight-fitting motorcycle jacket and absence of a shirt, the "Rod Stewart" haircut, that he "was ready to go out and party hard," and that "he'd be happy to share his views." In the privileged position as reader, one is tempted to dismiss Mehring's stoner-esque disquisition in much the same way that the likes of Alan Sokal and Jean Bricmont dismiss a sampling of postmodern philosophy in their highly influential work, *Fashionable Nonsense: Postmodern Intellectuals' Abuse of Science*, by asking "what philosophical function can be fulfilled by this avalanche of ill-digested scientific (and pseudo-scientific) jargon?" (155). Sokal and Bricmont argue that "[w]hen concepts from mathematics or physics are invoked in another domain of study, some argument ought to be given to justify their relevance" (9).[1]

One might object that Mehring's exposition does not qualify as a "domain of study" in any strict sense, since he would not seem to be speaking from a position of authority; the authority granted by a degree or any other form of institutional recognition. Nor does Mehring justify his invocation of string theory, apart from the implicit license he might be said to enjoy as someone attempting the ultimately unverifiable; that is, a coherent synthesis of cosmology with eschatology. Sokal and Bricmont—and tacitly, Chapman—undoubtedly would categorize Mehring's efforts as yet another example of the gross misappropriation and distortion of scientific knowledge by non-scientific discourse. Here an "average Joe" has consumed and is haphazardly regurgitating a mélange of scientific buzzwords, which happen, in this case, to hinge on this "new theory"—"the eleventh dimension."

Chapman ends his article on this passage. As a journalist committed to an impartial reportorial stance, he does not allow himself the liberty of pontificating on such heady topics, let alone asserting an emphatic opinion. Yet he and his editors permit a "testifier" to do so. One might even argue that Chapman is allowing Mehring to speak for him as a kind of proxy. For, in a culture whose epistemologies are increasingly fragmented into highly discrete specializations, the only narrators reckless enough to arrogate the authority required to synthesize knowledge into a totality are the clown, the artist, and the fool. As discussed in the previous chapter, when physicists claim that string theory is a theory of everything, they mean something specific to what the discipline accepts as legitimate problems to be solved within the stringently defined mathematical formalisms and experimentally validated evidence of the field. But, as in this example, when popular discourse evokes string theory as a theory of everything, that everything often grows to include much more than the "puzzle-solving" that Thomas Kuhn takes to be the core of theoretical physics praxis. Subject to alternate constraints, an imaginative treatment of a theory of everything is free to hop disciplinary boundaries, much in the same way that Mehring would have us "just somehow hop over to the eleventh dimension, and hop from universe to universe to universe forever inside the eleventh dimension."

I will return to Mehring's disquisition at the end of the chapter, but here I want to introduce a term that offers a useful, if not pivotal, method for deciphering texts that evoke string theory. Much has been made of the supposed schism in contemporary culture between science and art, the so-called Two Cultures, named thus by C. P. Snow in his highly influential lecture of 1959. The conventional realist account sets art and science in opposition as a binary. On the one hand, science cloaks itself in the rigors of logic and empiricism, while on the other, the arts make do with the less rigorous modes of rhetoric and mimesis. In *Picturing Science, Producing Art*, Caroline A. Jones and Peter Galison argue that in this account

> like all binaries, art and science needed to be yoked together (yet held apart) in order to accrue the strengths of their polar positions: soft versus hard, intuitive versus analytical, inductive versus deductive, visual versus logical, random versus systematic, autonomous versus collaborative . . . The binary production of knowledge (the bifurcation of practices) was equally simple: art invented, science discovered. (2)

Implicit in this paradigm is the conviction that translation of scientific concepts from technical discourse to non-technical discourse necessarily involves, at the very least, oversimplification, and more often than not, the kinds of distortions and malapropisms to which Sokal and Bricmont so vociferously object.

In *The Cosmic Web*, N. Katherine Hayles argues that certain forms of literature are "an imaginative response to complexities and ambiguities that are implicit in [scientific] models but that are often not explicitly recognized" (10). In keeping with Hayles's formulation, it will be useful to assume that the imagination functions as the primary vehicle of exchange between the two discursive cultures; that is, between the professional and the literary. But Hayles's statement suffers from a crucial elision, which, in a certain sense, is complicit with the binarism endemic to the Two Cultures paradigm. If on the one hand, string theory offers up a body of scientific knowledge expressed exclusively in terms of mathematical formalisms, while on the other hand, a literary text's "imaginative response" occurs within a distinct, non-mathematical language, then, assuming that, in general, literary authors are not professionally trained string theorists, a third discourse must mediate the exchange between the technical and the literary.

What I propose is that this intermediary be called a *scientific imaginary.* In effect, this is a study in intertextuality, and specifically, of the unidirectional translation of string theory as a scientific imaginary from one discursive domain—the technical, eventually to another—the literary. Although every particular text might be said to express its own uniquely specific version of string theory as an imaginary, it is worth exploring the extent to which that imaginary might be independent of specific texts. In this sense, string theory as a scientific imaginary may or may not retain some of its shape, feel, and internal structure as it migrates from one text to another, from one discourse to another. Out of this translation, a pattern of imaginative continuity may have important epistemological—and ideological—implications for string theory as a dominant scientific practice within a contemporary globalized culture.

Before considering which particular texts warrant attention, it is important to emphasize that it is insufficient to assume that only one imaginative leap, so to speak, bridges the chasm between technical and literary discourses. A crucial aspect of this investigation will be to consider the extent to which string theorists employ an imaginary within technical discourse itself. Specifically, the next chapter explores where and how key string theoretical images arise in the technical discourse of string theory and what roles these core images play within an overarching imaginary. Furthermore, the popular and literary texts considered in later chapters are not, by and large, direct imaginative responses to string theory as it is expressed in technical discourse, but most often, to its formulation within popularizations, which are accounts of string theory, usually authored by string theorists, and specifically addressed to a lay audience.

A complex technical imaginary is not what writers—of science fiction, for example, except perhaps in certain rare cases—directly consume and subsequently reproduce. String theorists undoubtedly would claim that mathematical arguments are the centerpiece, the principle content, of technical discourse. Whatever imagery technical articles might evoke, whether within the expository prose that surrounds mathematical argument or via graphic illustrations, they would insist plays a secondary, supplemental role within the discourse. Yet in string theory popularizations, largely free of mathematical content, the images implicit in the expression of scientific ideas take precedence as core content. This is a seeming contradiction that deserves investigation. If mathematical arguments contain exclusively the "truth" of string theory, what exactly do popularizations convey, in light of the relative absence of those mathematical arguments? The answer would seem to be: an imaginary. This, in turn, prompts the question: what is at stake in one particularly fraught intertextual transaction—specifically, that of string theory images that migrate from popularizations to literary texts and the subsequent transformation of a given imaginary of which those images play an integral part?

Authors of popular and literary texts must necessarily consume a version of string theory before they can reproduce it. There are isolated examples where authors mention in the paratextual material framing their texts the specific sources of their borrowings from string theory. But even such specific sourcings of ideas from string theory presuppose a certain cultural currency of string theory as a whole. Consumers of scientific ideas are, in a sense, primed to focus on particular theories that already enjoy a certain level of wider cultural circulation. Bruce Clarke and Linda Dalrymple Henderson describe this process as typically proceeding as follows: "In the case of the historical sequence from the nineteenth-century sciences . . . to the twentieth-century sciences . . . , experimental and theoretical developments have fed into technological innovations, which in turn have been commoditized and distributed as cultural practices throughout modern society" (1–2). The industrial production of electricity, for example, could be said to have precipitated this process for the theory of electromagnetism. Quantum theory, in turn, gained currency through the proliferation of computers—among a host of technological products—but especially through its connection in the popular imagination with the atomic bomb.

But other theories, such as Einstein's theory of special relativity, would seem to lack obvious technological applications.[2] In such cases, where the science remains predominantly theoretical—as with string theory—what becomes commoditized and distributed is not technology per se, in the form of industrial products or gadgetry, but the theoretical "ideas" themselves. These ideas, divorced from their mathematical contexts, take the form of reiterating images, conceits, and anecdotes, as with Mehring's notion of "the eleventh dimension." They form a kind *doxa*—a set of stock ideas, images, anecdotes, and scenarios that constitute an imaginary specific to that particular scientific theory. For example, nearly every popular account of quantum theory includes its own version of the double-slit experiment, the image of the wave/particle paradox, and the "Schrödinger's cat" thought experiment, among many others.[3]

Graham Allen argues that, in general, texts are never autonomous; they necessarily involve a certain intertextuality. Every text "sets going a plurality of meanings but is also woven out of numerous discourses and spun from already existing meaning. The text's plurality is neither wholly an 'inside' nor an 'outside,' since the text itself is not a unified, isolated object upon which an 'inside' and an 'outside' can be fixed" (67). The string theory technical articles found within the *hep-th* section of the lanl.arXiv.org e-Print Archive[4] form an important part of the technical discourse intertext, which in turn, overlaps and informs the intertext of string theory popularizations—both monographs and mass-market journal and magazine articles—that consequently feed into the intertext of fiction, plays, or poetry that engages, however indirectly, with string theory. From this perspective, the myriad, overlapping intertext functions as a vehicle for the circulation, perpetuation, and potential complication of string theory as an imaginative *doxa*.

The Imaginary in a Scientific Imaginary

What, then, is a scientific imaginary? I am adapting the expression from Michèle Le Doeuff. In *The Philosophical Imaginary*, Le Doeuff describes a fundamental dichotomy in philosophical discourse, one that privileges what she calls "concept" over "image." She argues that philosophical discourse traditionally views the image as superfluous and attributes it to one of two sources: to "infantile or primitive thought" or to "adaptation" for "didactic" purposes. Together imagery and knowledge form a "common system." She writes that "between these two terms there is a play of feedbacks which maintains the particular regime of the discursive formation. Philosophical texts offer images through which subjectivity can be structured and given a marking which is that of the corporate body" (5).[5] In her reading, images are didactic, primitive, or fanciful. They are associated with subjectivity, affect, and the "corporate body." The concept, on the other hand, possesses an epistemologically "pure" truth value; it is objective, and thus free of affect. However, philosophical discourse also conscripts the image to ground (surreptitiously, Le Doeuff suggests) its claims of objectivity on the subjective body through structured affect. By situating image adjacent to concept, philosophical discourse betrays a certain dynamic tension between the two, where image problematizes concept while ostensibly elucidating it for didactic or ludic purposes.

Le Doeuff sees the accumulation of imagery within a philosophical text as forming a whole, a whole that both exhibits its own internal structure and relates to the conceptual content of the text through juxtaposition.

> Images generally need to be decoded before one can relate their meaning to the thought made explicit in the text, in order afterwards to rein-

> troduce into the discourse the question which the image both resolves and helps to evade. But this reinsertion allows the hypothesis of a converse: if the images of philosophical texts are so functional, so organic in their very dysfunctionality, might we not guess that they are made to measure, that there is not just an imaginary *in* philosophy but a properly philosophical imaginary . . . ? (4, emphasis in original)

In keeping with Le Doeuff's formulation, I define an imaginary as a complex of images. It is a complex in that it comprises an arrangement with a structure where the internal relationships between the images exhibit an intricacy that allows for a freedom of signification among the images while simultaneously imposing certain constraints on that freedom. In this sense, an imaginary is a space of possibility, plausibility, and impossibility. An imaginary's structure implies a certain logic of organization whereby some images are central, others peripheral, and their placement and patterns of interaction display certain regularities.

It is worth noting here that Le Doeuff is addressing primarily an Enlightenment philosophical tradition whose careful separation of concept from image is very much in keeping with the kind of purification that scientific realism insists upon.[6] Scientific realism is relevant to this study of string theory in that physicists, in general, subscribe to its fundamental position—that mathematical arguments verified by experiment represent an objective reality. This will become more apparent in subsequent chapters when I examine instances in technical discourse and popularizations where the authors make a point of expressing their opinions on the status of string theory with respect to an objective world. It is, in effect, a rather small leap to speak of a scientific imaginary, in the realist tradition, in a manner consistent with Le Doeuff's critique of Enlightenment philosophy.

In the kind of philosophy that concerns Le Doeuff, imagination contrasts with the real; it is a fanciful or fantastic place of possibility freed categorically from the limiting constraints imposed by reality. The imaginary, that which the faculty of the imagination produces, is thus understood as a place of pure novelty; it is immaterial, and as such, utterly unmoored from the objective world. Within the imaginary, the images that constitute it function as what Gaston Bachelard calls, playing off the word's etymology, an "idealized double" (*Reverie* 176). Like scientific realists, Bachelard is interested in isolating the imaginative component from science proper by purifying scientific truth of any contaminating imaginative content. Scientific realists, drawing on the tradition of logical positivism, effect this purification by making a distinction in scientific praxis between a context of discovery and a context of justification (Reichenbach 231). An imaginary—and its concomitant freedom of movement and expression—finds its proper home within the context of discovery, which they associate with creativity. On the other hand, the context of justification requires a conceptual content purified of any imaginative contamination, of any subjectivity or affect. Le Doeuff argues that "Bachelard . . . has offered analyses of the imaginary component within scientific work, whose final aim is to extradite an element judged alien and undesirable, and assign it a residence *elsewhere*" (2, emphasis in original).

While scientific realists concern themselves primarily with the context of justification in scientific praxis, where, they contend, its truth-value ultimately resides, Bachelard, especially in his later writings, privileges what he sees as a "material imagination"—a pure source of creativity.[7] Where the context of justification—with respect to string theory, principally mathematical argument—privileges abstractions that make claim to a universality, as Bachelard argues in *The Poetics of Space*, the material imagination is akin to a "space that has been seized upon by the imagination" (xxxvi). As such, it "cannot remain an indifferent space subject to the measures and estimates of the surveyor. It has to be lived in, not in its positivity, but with all the partiality of the imagination" (xxxvi). It is not an abstract space but an inhabited place.

In *The Book of Skin*, Steven Connor summarizes Bachelard's material imagination.

> There is no way of imagining the nature of the material world which does not draw on and operate in terms of that material world. So imagination is itself always implicated in the world that it attempts to imagine . . . Is it possible not to imagine such things in a muscular fashion, in terms of the resistance or release that we would feel in encountering them, in other words in terms of the theories of the nature of such material forms that are embodied in our habitual or learned comportments towards them and our likely or possible bodily interactions with them? The image in each case would be much more than something merely seen; it would be, it must be, not only image but also usage. So the phrase "material imagination" must signify the materiality of imagining as well as the imagination of the material. (40–41)

The world we imagine, however abstracted and universalized in a scientific theory such as string theory, must bear with it an intimacy that stems from embodied interactions with that imagined world. It is this relationship between abstraction and, as Connor puts it, "our habitual or learned comportments" that constitutes a "materiality of imagining." Such a materiality of the imagined suggests a complication of not only concept and image, but also the simple dichotomy between the abstract and the concrete. In a sense, the pure work of theorizing—mathematical argument as scientific praxis—becomes contaminated by the culturally conditioned bodies that do this work.

A by-product of this separation of image from concept is a perpetuation of the cultural chasm between the sciences and the arts. If one accepts the strictures on truth-value imposed by scientific realism, then, with regard to string theory, one is forced to concede its objectivity resides exclusively in its mathematical formalism. This concession, in turn, prompts the question of how to interpret the rest of string theory's content: the exposition—rife with imagery—that surrounds mathematical argument. Michel Serres suggests in *Conversations on Science, Culture, and Time* that Bachelard "consummated the rupture . . . between science and the humanities—perceiving on the one side a spirit of . . . working and, on the other, a material imagination that sleeps, dreams, and ponders" (31). The ultimate consequence of this extreme polarization of image from concept, Serres argues, is that science becomes "founded on itself and, therefore, has no need of external philosophy; it contains its own endo-epistemology" (128).

But again, even with a cursory examination of string theory's technical discourse as a collection of textual artifacts, one observes a close proximity between image and concept. Exposition—written in ordinary language—frames, links, and sustains mathematical praxis. By a scientific realist's own definition, that exposition lacks rigor—a positive truth-value. As such, the rupture between image and concept would seem to be more of a hairline fracture than a gaping chasm. As soon as string theorists "expose" the formalism through non-technical exposition, they invoke an imaginary. The gap between concept and image, then, is the space on the page between a calculation and its explication. Epistemologically denigrated, the "dreams" of string theory as material imagination become displaced into supplemental and ostensibly superfluous exposition. In effect, exposition must do the dirty work that, while necessary, concept itself cannot acknowledge.

Can exposition be objective in the way that string theorists claim mathematics is when it is both logically consistent and confirms experimental data? In other words, is it possible to communicate scientific knowledge through expository prose in a way that would be conceptually pure; that is, free from contamination by an imaginary? Exposition may exhibit logical consistency, but string theorists would argue that it lacks the precision—

and conciseness—necessary for rigorous descriptions of subatomic scales. A major mechanism in expository prose—that which contributes substantially to the production of meaning, even in highly abstract contexts—is that it employs, to some varying degree and intensity, the concrete; that is, representations of the embodied world of everyday objects, events, and relations.

The analysis that follows relies heavily on the work of the cognitive linguist George Lakoff and his various collaborators, in large part, because not only does Lakoff stress the centrality of the image to language, his writing offers a systematic, empirically grounded, and highly productive taxonomy of core images and their dynamics within natural language. In *Philosophy in the Flesh*, Lakoff and Mark Johnson point out that "first generation" cognitive linguists of the early twentieth century claimed that the language of formal logic is an exception, in that images are generally stripped away and replaced with "pure" symbols (75–78). Formal logic—and an epistemology that equates mathematics with formal logic—thus privileges the relationships among the symbols, rather than any correspondence between symbols and embodied human experience. Beginning in the 1960s, "second generation" cognitive linguistics, on the other hand, has emphasized the centrality of the imagination to linguistic expression (77). Lakoff and Johnson observe that "words can designate portions of conventional mental images"; "mental images do not vary widely from person to person [and] conventional mental images are shared across a large proportion of the speakers of a language"[8]; a "significant part of a cultural knowledge takes the form of conventional images and knowledge about those images"; and finally, the "meaning of the whole is not a simple function of the meanings of the parts" (69). They emphasize that, between these images, the "relationship is complex"; it consists of the "linguistic expression of the image plus knowledge about the image plus one or more [inferential] mappings" (69).

In keeping with first generation cognitive linguistics, in *The Rule of Metaphor*, Paul Ricoeur argues that it is indeed possible to isolate the conceptual content of prose from its imaginative content.

> The signifying aim of the concept works free of interpretations, schematizations, and imaginative illustrations only if a horizon of constitution is given in advance, the horizon of speculative *logos*. By reason of this opening of horizon, the concept becomes capable of functioning semantically in terms of the configurational properties of the space in which it is inscribed. . . . Because it forms a system, the conceptual order

> is able to free itself from the play of double meaning and hence from the semantic dynamism characteristic of the metaphorical order. (302, emphasis in original)

What Ricoeur describes here—expression where "concept becomes capable of functioning semantically in terms of the configurational properties of the space in which it is inscribed"—is epitomized by the language of formal logic. To reiterate, such language privileges internal relationships within the system over the "double meaning" of images. Ricoeur believes that exposition may "free" itself from "the play of double meaning" through precisely these kinds of "configurational properties." Tellingly, though, he justifies the efficacy of this "conceptual order" by means of an image; namely, that of an "opening of horizon."

While Ricoeur argues, rather contradictorily, that concept in exposition can liberate itself from imagery, he also claims that imagery may still be useful. Images, he writes, "are not things at all; rather, they introduce a new language, like a dialect or idiom, in which the original is described without being constructed"; an "imaginary medium is . . . nothing more than a mnemonic device for grasping mathematical relationships" (241). Such a definition formalizes the commonsense understanding of the relationship between a word and the object to which it refers. There is the word *string* and the mathematical object to which it refers—an object which finds its meaning within a system of relationships codified through mathematical formalisms. According to Ricoeur, a word such as *string* would not construct the original per se, but merely describe it for a didactic or illustrative purpose. Ricoeur's conception—or rather, way of imagining concept—ignores the cognitive work that makes an object recognizable as an object; or in other words, the way in which we as embodied, cognizing, acculturated agents interact with objects in the world.

Ricoeur's insistence on the discrete freedom of concept from the corrupting "double meanings" of image echoes the logical positivist sorting of the context of discovery from the context of justification. He states: "The important thing is not that one has something to view mentally, but that one can operate on an object that on the one hand is better known and in this sense more familiar, and on the other hand is full of implications and in this sense rich at the level of hypotheses" (241). Here, again, Ricoeur acknowledges the utility of an imaginary; that it draws upon images from everyday experience, which are thus "better known" and "more familiar." But the imaginary's utility once more lies in its capacity to generate "implications" and "hypotheses"; in short, as a form of play. Consistent with

Ricoeur's formulation, scientific realism associates the conceptual with work, and the imaginary with school and play. Since the conceptual work to which Ricoeur refers takes place, by the theorists' own admission, in the mathematics, the exposition that frames this work is accordingly not subject to the same stringencies of form; it is not couched in the language of formal logic.[9] In this sense, the exposition reflects a less concerted effort to purify it of contaminating imagery. A reader is meant to take seriously only the conceptual. Whatever imagery is latent in the abstractions that arise from conceptual work, one is simply expected to ignore or dismiss out of hand. Le Doeuff describes this expectation implicit within conceptualizing discourse, whether philosophical or scientific, as a "dogmatization." She writes that "images are the means by which every philosopher can engage in straightforward dogmatization, and decree a 'that's the way it is' without fear of counter-argument, since it is understood that a good reader will by-pass such 'illustrations'—a conviction which enables the image to do its work all the more effectively" (12).

With respect to string theory discourse, the images employed function, on the one hand, as doubles. They point to and evoke particular mathematical formalisms—as placeholders or cynosures. On the other hand, such an imaginary is material in a manner distinct from its function as a double. An imaginary, in this sense, performs a paradoxical double-double. Firstly, it points to and doubles for the objectivity of the mathematical formalisms, the conceptual content of a given technical article. As Bachelard declares in *The New Scientific Spirit*, "there is more to mathematics than formal structures, and . . . every pure idea is accompanied by an imagined application" (4). Secondly, an imaginary also evokes a distinct and autonomous materiality—the materiality of an image's associations and connotations with reference to the realm of human-scale experience. This is yet another doubling, what Bachelard calls doing "duty for reality" (4).

An imaginary constitutes an intermediary between a world "out there" and human creativity. It mediates interventions into a causal structure that exists independently of human desire, perception, or cognition. It functions, in this sense, as more than a cynosure, but rather, as a template. A template has a dual nature: on the one hand, it represents a self-consistent whole with internal relations amongst mutually constituting objects, while on the other hand, it refers to something outside itself. It serves as a guide for intervention into an objective world. As a relatively self-consistent and autonomous whole, like a template, an imaginary has the potential to be medium-independent. Templates can be reconfigured and redeployed in various media whilst still preserving essential object-identities and internal relations. This is not to suggest that templates are immutable—changes in configuration may transform irrevocably the nature of an imaginary to the extent that it would no longer be recognizable from an original. One could argue that transformations of an imaginary fall along a spectrum where one pole represents an exact reproduction and the contrasting pole, an utterly unrecognizable transmutation. The point here is that an imaginary, while being heterogeneous in constituent parts and relations, may also exhibit a certain plasticity that does not necessarily threaten the integrity of its cohesion.

Since images bear with them material, cultural, and historically contingent aspects, an imaginary cannot be said to be categorically novel in that it is absolutely detached from the real—or in other words, the already known. Rather, an imaginary represents a novel recombination of constituent elements, drawn from the familiar realm of experience. It is populated with images accessible to imagined human hands, eyes, and ears, however abstracted. Linguists Gilles Fauconnier and Mark Turner describe the work of the imagination thus: "We divide the world up into entities at human scale so that we can manipulate them in human lives, and this division of the world is an imaginative achievement" (8).[10] Fauconnier and Turner also suggest that one of the benefits of an imaginary "is its ability to provide compressions to human scale of diffuse arrays of events" (30).[11] If one conceives of mathematics as a kind of abstracted tool for conducting precise and repeatable interventions into the causal structure of an objective world, then a corresponding imaginary serves it as an ergonomics: a good tool must not only fit the material one is working with, it must also fit the hand well. Compressions and expansions to human scales through an imaginary allow an imaginary to have a certain familiarity and comprehensibility. The stability of an imaginary's comprehensibility is predicated on the familiarity of the images that constitute it. Yet the imaginary grows in complexity inasmuch as its images possess a wide array of characteristics; the subtleties of their connotations within the overarching structure multiply.

Bachelard describes scientific progress in terms of a series of stages: its conceptual content evolves from a concrete stage to a concrete-abstract stage, and ultimately to a fully abstract stage (*Formation* 20). The mathematical formalism of string theory, in this interpretation, would represent a fully mature form of scientific praxis—a thoroughly rigorous abstraction. But a string theory imaginary does not conform necessarily to this neat progression. It may draw upon the full repository of images that exist in a culture or multiple cultures—whether fully concrete, abstract, or some shade in between. Furthermore, as a complex, images with radically disparate qualities may well be juxtaposed. In effect, the structural logic of an imaginary—its composition and internal relationships—is not necessarily bound by the same constraints as the mathematics to which it originally corresponded. A schema that accompanies a given image may also constrain it—it represents an inferential structure implicit in the image based on common experience; that is, the experience of interacting with an object through a body situated in space and time.

One example of an image schema is a container. All containers possess an identical structure: an inside, an outside, and a boundary. A person learns to conceptualize, imagine, and manipulate containers through interactions by means of his or her hands and eyes. Specific experiences of containers become gradually generalized into an abstracted, prototypical container, which can be represented by a single image. There are also historically and culturally contingent containers: the gourd, the vase, the silo, the oil tanker, the mind, the universe, etc. Some cultures would recognize the mind or the universe as a container, others would not; they would also imagine the inside, outside, and boundaries somewhat differently. Some containers are closed, for instance, and some are open—in different ways. Nevertheless, the generic image of a container is relatively stable throughout history and across cultures—with its fundamental structure of inside, outside, and boundary.[12] In effect, what is possible in an imaginary is bound by the inferential structures of the images that populate it.

The internal structure of an imaginary also works to substantiate the images it subsumes. Patterns of interaction contribute to a sense that the images involved have substance. The images that comprise the imaginary have substance in that they possess a certain context-specific objectivity—these image-objects not only can be apprehended cognitively, but also have a context-specific solidity or even reality. More than mere shadows or doubles, within an imaginary, images become manifest within a space where imagined human agents may interact with them. Human agents are able to apprehend the images as objects, act upon and manipulate them. The images-as-objects may also, in turn, exercise their own form of agency—a kind of animation of material objects. I use the term *agent* in order to emphasize the sense of interaction that takes place within an imaginary. As an intermediary between human-scale embodied experience and remote realms, an imaginary provides a vehicle—and justification—for intervention.

In *The Mangle of Practice*, Andrew Pickering makes precisely this argument when he claims that "scientists, as human agents, maneuver in a field of material agency, constructing machines that, as I shall say, variously capture, seduce, download, recruit, enroll, or materialize that agency, taming and domesticating it, putting it at our service, often in the accomplishment of tasks that are simply beyond the capacities of naked human minds and bodies, individually or collectively" (7). As a sociologist, Pickering is concerned with interrogating scientific praxis directly. Were he speaking of string theory as a science, rather than "constructing machines," he would presumably speak of its construction of mathematical formalisms as abstracted "machines." It is the experimental physicists who work with machines, the colliders and accelerators at labs such as CERN, while string theorists currently stake their claims to realism on the internal consistency of their mathematical arguments and their capacity to reproduce the standard model. I want to stress that I am making no claims regarding string theory's status as legitimate science. Nevertheless, the imaginary manifest in the exposition surrounding string theory mathematical argument does, in fact, enable the kinds of interactions between human and non-human, material agencies that concern Pickering, such that "human agency and captured material agency are . . . constitutively intertwined" (17). Furthermore, this encounter between imagined human agents and material objects that also exhibit agency contributes to the reader's impression that the images that constitute a given imaginary indeed have substance and, as such, constitute a "deeper" reality.

If, as Connor contends, "there is no way of imagining the nature of the material world which does not draw on and operate in terms of that material world," then for an image to have substance, first of all, it has to be recognized as an object, distinct from a background. Accordingly, it must be imagined as perceivable—by at least one of the five senses. In the case of touch, an image with substance must have heft—it is imagined as graspable, embraceable. How can a string, then, be said to have substance when it is made of energy and is only accessible indirectly through mathematics and particle accelerators? If, in the future, experimentalists were able to construct a collider powerful enough to probe the Planck scale, the evidence confirming the string's existence would still only be circumstantial—data from the computers linked to the collider's detectors. Even the cloud chambers of first generation quantum mechanics experiments did not capture the traces of the subatomic particles that resulted from the collisions, but rather the trace of the water vapor molecules hit by those particles. In high energy physics, substance is a matter of inference and abstraction. Fundamental objects such as the quark and the string have substance only in the context of the imaginary that accompanies the formalism and experimentation. These image-objects, impossibly remote from human-scale apprehension, are seen and grasped by what I suggest are, in a sense, *homunculi*: imagined human agents—the mind's eye and hand—projected into the imaginary at a scale appropriate for embodied interaction with the imaginary's occupants. In an imaginary, human doubles interact with abstract theoretical objects made "substantial" through that very interaction. In this sense, an imaginary both doubles and extends the boundaries of the world we are accustomed to perceive, understand, and operate within. Yet this is not to suggest that an imaginary resembles or reproduces a human-scale world with complete faith or accuracy of taxonomy, interrelation, or proportion. It is not realist in the way the imaginary of a circumstantial realist novel is understood to be.

An imaginary such as that of string theory also implies a totality in that it functions metonymically. An imaginary gestures toward a coherent whole (even as it may not succeed in fulfilling its intention to be coherent). As a totality—or totalizing gesture—an imaginary engenders a world order. The world is the totality—that amalgam of images and their rules of interaction that, in a reductionist strategy such as that of string theory (which, in its crudest articulation, declares that the world is made of strings)—attempts to extrapolate the whole from its parts. But a world order also, importantly, contains an order; that is, a regular structure that is taken to define and thus be the essence of that world. And our capacity to understand that world order necessarily implies a human agency that is capable of engaging with it. A world order is synoptic—a definition of order, formed through a consensus of human agencies, gathers together and constitutes a whole. The two elements are imbricated—a world and those who would re-cognize it. I use the plural here to emphasize the fact that string theory works by consensus.

With respect to a string theory imaginary, then, one could replace the term *world order* with *cosmic order*. The advantage of speaking of a string theory imaginary as a cosmic order is simply that the word *cosmos* yields a more expansive connotation than *world*. Where world often implies solely planet Earth apprehended on human scales or global scales (for example, in the term *world peace*), cosmos clearly designates the universe in all its vastness and totality. Yet the term *cosmic order*, like world order, also conjoins the outside with the inside—the comprehension of its order, with, importantly, the communication of that order, where communication necessarily implies a discursive community. As such, with respect to cosmic order, a community and the discourse that expresses the order of the world (and that community's place within that order) are mutually constituting. An imaginary that serves as a cosmic order mediates the interaction between cosmos, the so-called "objective" world out there in its fullest range from microscopic to macroscopic, and culture—the social agencies and ordering practices of a human community. A cosmic order is an imaginary situated between culture and cosmos: it implies a signifier, the cosmos, a signified, its meaningful order, and one (or many) doing the signifying. In the case of string theory technical discourse, this would be the theorist and his or her colleagues, collectively self-recognized as a professional community.

In *Beamtimes and Lifetimes*, Sharon Traweek defines culture as "a group's shared set of meanings, its implicit and explicit messages, encoded in social action, about how to interpret experience" (7). The reverse is also the case: a culture is a shared set of social practices encoded in a symbolic structure. Traweek's emphasis is sociological. She is referring to social action within the physics community: theorists doing calculations on scratchpads, whiteboards, or computers; theorists attending conferences, conversing with other theorists, publishing papers, exchanging emails, etc. With respect to an imaginary, what is primarily relevant are not the social practices of string theorists in the doing of physics, but rather, representations of a wide range of social practices that occur within the imaginary—human-like agents acting upon objects, objects acting upon agents, and objects acting upon each other. Within a string theory imaginary, then, representations of social action generate interpretation just as much as interpretation, that "shared set of meanings," generates social action.

From this perspective, the contrast between cosmos and culture becomes more a matter of emphasis. Within an imaginary, the objective world becomes a projection of a community's self-regulated structure of social actions—and vice versa. This coupling of cosmos with culture is an idea originally given currency by the sociologist Émile Durkheim. Traweek, echoing Durkheim, writes that "a culture's cosmology—its ideas about space and time and its explanation for the world—is reflected in the domain of social actions. In other words, ideas about time and space structure social relations, and the spatial and temporal patterns of human activity correspond to people's concepts of time and space" (157).[13] A culture's notions of the world it inhabits inform its social practices while its social practices shape its notions of the world. In the context of an imaginary, to speak of culture is to focus on what Michel Serres calls "our relations among ourselves," while to evoke the cosmos is to focus on "our rapport with things" (*Conversations* 141). Serres uses the example of the postwar space program to highlight this imbrication of culture and cosmos: "Every technology transforms our rapport with things (the rocket takes off for the stratosphere) and, at the same time, our relations among ourselves (the rocket ensures publicity for the nations that launch it)" (141). He adds: "This object, which we thought simply brought us into relationship with the stars, also brings us into relationship among ourselves" (148). One may speak of a continuum between "scientific" and "cultural" imaginaries, where an emphasis of orientation "out there" or "among us" determines the imaginary's status and function. As mentioned previously, with respect to string theory, the objects that will concern us are not technological, like the Apollo Saturn V rocket, but theoretical.

In *The Fourth Dimension and Non-Euclidean Geometry in Modern Art*, Linda Henderson investigates one highly influential case of such a coupling of an "out there" with an "among us." Henderson shows how certain early twentieth-century artists, often associated with the modernist movement, equated non-Euclidean geometry, and in particular, the notion of the "fourth dimension," with "the rejection of tradition and even with revolution" to the extent that this equating of hyperspace with sociopolitical concerns became what she calls a "hyperspace philosophy," whose prominent champions included C. H. Hinton and H. G. Wells (17, 25). She writes that those who embraced hyperspace philosophy "believe firmly in the reality of a fourth dimension of space, yet tend to oppose any form of positivism that requires empirical proof of its existence," and that the "underlying theme is generally that the answer to the evils of positivism and materialism is for man to develop his powers of intuition, in order to 'perceive' the fourth dimension of our world, the true reality" (25).

In keeping with Serres and Henderson, I want to briefly explore three other examples of cultural readings that reflect this close coupling of a cosmic order to representations of social practice in order to explore the implications of their variations. The first comes from Bertrand Russell's *The ABC of Relativity*. In this passage, Russell draws a link between the physics concept of force and politics.

> If people were to learn to conceive the world in the new way, without the old notion of "force," it would alter not only their physical imagination, but probably also their morals and politics. . . . In the Newtonian theory of the solar system, the sun seems like a monarch whose behests the planets have to obey. In the Einsteinian world there is more individualism and less government than in the Newtonian. (2)

This quote nicely illustrates how an imaginary functions. Both the "Newtonian theory" and Einstein's special relativity are formalisms expressed exclusively in the language of mathematics. Russell would seem to be conflating an exposition of these theories—what he calls "their physical imagination" with the theories themselves. This, in turn, allows for an easy imaginative leap to the discursive domains of "morals and politics." Sokal and Bricmont would certainly argue that such an imaginative leap betrays an aggressive and perhaps inappropriate adaptation—a distortion. Russell, though a self-professed socialist, would seem to be hijacking a scientific theory in the service of a libertarian politic, one that valorizes individualism and decentralized authority. Yet this particular example is notably simplistic.

The second example comes from Mary Midgley in her work *Science and Poetry*. "The social development of individualism increased the symbolic appeal of physical atomism, while the practical successes of physical atomism made social individualism look scientific" (10). Like Russell, she is speaking of the relationship between Newtonian cosmology and "social individualism." In this cultural configuration, the image of the atom and the image of the individual become mutually reinforcing. But rather than co-opting a scientific imaginary to promote a political view, Midgley calls attention to the coupling of that imaginary to the culture in which it flourishes. There is a sense that Midgley is aware of the fracture between formal Newtonian theory and a scientific imaginary predicated on it. Unlike a realist perspective that insists upon an absolute distinction between the conceptual content of scientific practice and the imaginative content of non-scientific discourse, Midgley here recognizes the amplifying feedback that a scientific imaginary can supply to a given cultural predilection.

Kirsten Shepherd-Barr provides the third and final example, from her book *Science on Stage*. Speaking of Galileo's theory of heliocentrism, Shepherd-Barr suggests that "the discoveries he made helped to bring about a fundamental change in how we view the world" (28). A self-evident truism like this has such universal currency it seems unassailable. Yet it belies a fundamental conflation akin to that of Russell, one that, on closer inspection, helps to further illustrate how a scientific imaginary functions. In spite of our complete conviction that the Earth revolves around the sun, we still bear direct witness to the sun revolving around the Earth every day. What we now do, though, is *imagine* the Earth moving around the sun, and mark that imaginary as the truth, the deeper reality. That truth bears the authority of consensus over many generations; it is a highly stable knowledge. Most adherents of the heliocentric theory cannot prove its veracity: they accept it as a matter of dogma. We defer to the specialists who are altogether capable of conducting (and have conducted, as a matter of historical record) experiments to verify its truth-value.[14]

In the heliocentric imaginary, the relationship of Earth to sun even makes use of the geocentricity of common experience. It employs a structure of correspondences between the images (the sun and the planets) and astronomic observation (with the aid of telescopes), but reverses the dynamic and the scale (Earth shrinks while sun expands). The heliocentric imaginary also draws upon reinforcing images of Earth-images revolving around sun-images, including, but by no means limited to, orreries, graphic illustrations, video simulations, or declarative statements like "the Earth revolves around the sun" espoused by authorities such as primary school science teachers. The one imaginary does not merely replace the other. We hold in our mind's eye the geocentric daily experience *and* the heliocentric imaginary contradicting while clarifying that experience. Our world has thus become all the more complex and collective, for now we depend on ever more specialists to reveal to us and each other, by means of an imaginary, the various realities—realities that a culture may order through a hierarchical image of surface and depth, the conceit of a "deeper reality" alluded to earlier. This world becomes populated with ever more intricate networks of sometimes complementary, sometimes contradictory, imaginaries (or imaginaries both contradictory and complementary simultaneously).[15] In effect, it is composed of a radically heterogeneous multiplicity.

An inclusive and general definition of an imaginary provides for the possibility that an imaginary would indeed incorporate images based on what a cultural consensus considers real objects and events, but it holds the potential to also recombine them into novel configurations along radically heterogeneous planes. Essentially, the imagination constitutes the novel through a recombination of the known and familiar. One might argue that, due to its scope and complexity, most changes to an imaginary are trivial and therefore do not generate novelty. In a physical theory such as string theory, the goal is not novelty for its own sake, regardless. According to the theorists, its goal is a novelty that both reproduces realistic physics as stipulated by quantum theory and general relativity *and* that gestures towards an empirically verifiable extension of its descriptive power beyond the range and precision of its two predecessors. String theory must be simultaneously novel and more real.

Any imaginary is constrained insomuch as the images that constitute it originate in the known world. Furthermore, because these images imply their own intrinsic inferential structures—image schemas—drawn from history, culture, and spatiotemporal embodiment, these constraints are not easily overcome. Every imaginary must contend with a host of cultural, historical, and spatiotemporal constraints that would infringe upon its paradigmatic liberty.[16] As a consequence, the known world becomes an overarching and encompassing intertext of concrete images, hybridized concrete-abstract images, and ostensibly pure abstractions—a cascade or tangle of images, schemas, ideas, and concepts that together comprise the sum total of a culture's knowledge of the world. In this imaginative milieu, the categorical distinction between abstract and concrete becomes nebulous. At best, one might define abstraction as the movement away from the specific to the universal. Yet our capacity to recognize the specific is predicated cognitively on a preexisting internalization of the universal. Given this state of affairs, a close reading of technical texts must sensitize itself to the complex relationship between exposition and the imaginary at its core, and the ostensive purity of mathematical formalism as abstraction. Furthermore, just as string theorists meticulously manipulate the intricate structures of mathematics, the imaginary that corresponds to and shadows the mathematics reflects its own intricate structure. That structure—the particular composition of images and their relationships—necessarily displays its own cache of ambiguities and ambivalences, some that echo the formalisms, some that do not.

Mathematical operations are highly precise and local: *this* calculation using one technique, *that* proof using another, etc. To stitch these local practices into a whole necessarily requires an imaginary. This proclivity to fill in the gaps between what are highly local scientific practices speaks to the significant role that an imaginary plays within a discourse that claims to be scientific. A scientific imaginary—a complex of images that references, with varying degrees of ambiguity, a self-professed scientific practice such as string theory—must necessarily betray a susceptibility to *scientism*. I want to clarify here what I mean by scientism, a word with strong ideological connotations. Originally, in the first decades of the twentieth century, scientism functioned as a descriptive term that was more or less synonymous with logical positivism. It engendered an optimistic anticipation that a strict execution of *the* scientific method would eventually lead to the full divulgence of nature's secrets.

That conception of science persists. In a 2002 column for *Scientific American*, Michael Shermer defines scientism as "a scientific worldview that encompasses natural explanations for all phenomena, eschews supernatural and paranormal speculations, and embraces empiricism and reason as the twin pillars of a philosophy of life appropriate for an Age of Science" (20). The implication here is that science, as a cultural practice, is monologic. Consider how many times one hears (or reads) in the media the following phrase: "Science has made a breakthrough . . . they've . . ." a multitude speaking with one voice from the same vantage point. Shermer's scientism proclaims that science, presumably of a monolithic civilization, ultimately unifies knowledge of the objective and objectifiable world into one coherent whole. Scientism tends to assume that there is only one empiricism, rather than a multitude of empiricisms—a farrago of methods for measuring and justifying that are, more often than not, specific to a given situation. It assumes that there is one monolithic reason, one categorical ur-reckoning, rather than a multitude of context-specific reasons. For Iain Cameron and David Edge, the motivation to espouse this form of scientism originates in the authority that the scientific community enjoys within Anglo-American culture; an authority predicated on "widely shared images and notions about . . . its beliefs and practices" (3). Scientism's proponents seek to "add weight to arguments they are advancing, or to practices they are promoting" by borrowing on such scientistic authority, which, in turn, reinforces and further consolidates that authority (3).

Bachelard is one of the first contemporary philosophers of science to call this perspective into question. "When one looks at science, what is immediately striking is that its oft-alleged unity has never been a stable condition, so that it is quite dangerous to assume a unitary epistemology" (*New* 14). Serres also takes this position when he writes that "reason makes use of concepts, under whose unities are sheltered multiplicities that are most often highly dispersed" (*Genesis* 3). But it is Paul Feyerabend in *Against Method* who argues most insistently that, as far as *the* scientific method is concerned, the sole dictum is that "anything goes." He declares that "the assumption of a single coherent world-view that underlies all science is either a metaphysical hypothesis trying to anticipate a future unity, or a pedagogical fake; or it is an attempt to show, by a judicious up- and downgrading of disciplines, that a synthesis has already been achieved" (14, 245).[17]

Paradoxically, such monologizing of the congeries that is scientific praxis generates what Bruce Clarke calls "diffusionist scientisms." Clarke suggests that they are "illegitimate offspring of science that cobble to an extrascientific object the cultural aura of science's own epistemological prestige, typically by an extension of scientific terminology, imagery, and/or methodology" (*Energy* 59). As "aberrant discourses," these "abuses bottom out in 'pseudoscience'—bogus representations or active misuses of scientific ideas" (60). "In sum," he writes, for contrarians, "as a repository for derivative and deformed conceptions *scientism* has typically been a term of bad intellectual repute" (60).

Yet, resonating with both Serres and Feyerabend, Clarke insists that "no cordon sanitaire actually keeps technoscience apart from its social causes and effects," in large part, because technoscience consists of "an epistemologically ambivalent and historically dynamic body of heuristic vehicles and visionary hunches embedded in heterogeneous yet interconnected cultural matrices" (60, 62).[18] Ultimately, he writes, scientism "is a dynamic and unpredictable irruption within the noise of technoscience, giving rise sometimes to dubious social or political doctrines, sometimes to striking productions of artistic and critical self-organization, sometimes to reorganizations of the entire cultural field" (65).

Scientism, whether positivist or pejorative, can only work through an imaginary. Without a scientific (or perhaps, scientistic) imaginary, any scientism would be incapable of sustaining the kinds of accretions and concatenations necessary for a grand unification. I want to stress, again, that this claim does not impinge upon the effectiveness of the scientific practices themselves. It does not question theoretical physics' access to an objective reality by means of reasons and methods, but rather, that in order to have local scientific practices cohere into a whole, one must necessarily resort to an imaginary. Pure concept, defined earlier as mathematical argument, alone will not serve.

Furthermore, a scientific imaginary such as that of string theory would ostensibly have a special, heightened authority, insomuch as it is presumed to be predicated on the general authority of science, in other words, on scientism as an epistemological stance. Skeptics, as discussed, use the term *scientism* derisively. It is meant to call attention to an almost religious faith in science's supposed omnipotence. But when string theorists take as self-evident their own endo-epistemology, they are, in effect, monopolizing authority to philosophize about string theory. Positivist scientism obviates the problem of multidisciplinarity—that specialization typifies scientific practice. In actuality, more often than not, the left hand of science does not know what the right hand is up to. An expert in one discipline has no more authority to assess the verity of another discipline than a lay person. As praxis, science is also fractured—neither microbiologists, neurochemists, solid state physicists, nor loop quantum gravity theorists is technically qualified to pass judgment on the value of a given piece of string theory. Often, a theorist who specializes in the applications of string theory to black hole cosmology may not be able to speak with any definitive authority on the subtleties of string field theory. Ultimately, in order to form judgments, specialists in one field only have access—assuming they refuse to sacrifice the years required to hone their knowledge in that other field—to expository accounts of other fields, and thus, to an imaginary. String theory, itself multiple and fractured, finds its place in a veritable Tower of Babel of scientific discourses where cross-fertilization and cohesion occur, in many crucial respects, by means of an imaginary.

The Science in a Scientific Imaginary

To reemphasize, as a scientific discourse, string theory consists of two basic components: an amalgam of mathematical formalisms and an imaginary that surrounds and sustains that system. The theorists themselves undoubtedly would insist that the actual scientific practice of string theory consists of the manipulation of mathematics—an assertion justified by scientific realism. Generally speaking, the professional culture of string theory endorses the maxim, attributed originally to Galileo, that the language of nature is best understood as mathematics.[19] Accordingly, the principal content of string theory—whatever truth-value it possesses—consists in its mathematical formalisms. As alluded to earlier, this is what scientific realists, adopting the term from the logical positivists, would call the context of justification, the proper field of scientific practice. The images or exposition that occur around the mathematics; for example, in a technical article—reside within the context of discovery. According to realism, images—and whatever imaginary they comprise—play a fundamentally didactic role in the practice of string theory as science. As a consequence, while theorists—in, for example, the popularizations they author for a lay audience—may very well concede that a given imaginary they present for string theory confounds commonsense intuition, that it lacks coherence, string theory as mathematical formalism must be entirely self-consistent.

Yet theorists have found it difficult to establish a definitive coherence within string theory as science proper. As explored in the first chapter, since its inception in the early 1970s, there have been numerous versions of string theory: initially, it was understood to be a theory of hadrons, then of bosons, then of bosons and fermions. A string theory that incorporated supersymmetry—known as *superstring theory*—reached maturity in the mid-1980s. Within superstring theory, five distinct variations have been afforded a certain credibility—Type I, Type IIA, Type IIB, Heterotic O(32), and Heterotic $E_8 \times E_8$. Each one of these five types can also be formulated in a multitude of ways, depending primarily on how a theorist compactifies the six extra spatial dimensions. Strategies of compactification include employing Calabi-Yau manifolds—of which there exists millions, if not billions—as well as orbifolds and orientifolds.[20] Others, such as S. James Gates Jr., have attempted to formulate superstring theories in four space-time dimensions.

With the emergence of M-theory in the mid-1990s, the multiplicity of string theories becomes even more compounded. As the first chapter described, M-theory suggests that a particular version of supergravity—a theory wholly distinct from string theory and one that, furthermore, possesses its own cornucopia of variations—represents one facet, along with the five previously mentioned superstring theories, of an overarching and inclusive "mother," "matrix," or "membrane" theory. The previous chapter also alluded to the fact that neither M-theory's creator, Ed Witten, nor the rest of the string theory community consider M-theory to be fully realized—it only sketches the limits of a coherent theory. Moreover, M-theory is not a theory of strings per se, but of branes. Strings become a special case of this more fundamental object.[21] More recently, string theorists have developed models that consider a cosmos where the extra spatial dimensions stipulated by superstring theory are not compact, but rather, infinitely expanded—giving way to various string theories where braneworlds exist within a metaverse, or what they have come to call a "bulk." Another braneworld theory that has garnered a great deal of attention recently is Landscape theory.[22]

The theorists themselves are hard pressed to articulate a compelling argument for the inclusion of this hodgepodge of theories, models, and configurations—all with varying degrees of distinctiveness—within some aegis called *string theory*. String theory as a moniker becomes largely a matter of convenience—and convention. What this heterogeneity suggests is that, while not science proper, as the string theorists themselves would define it, the imaginative component of the theory plays a significant role in its scientific practice—in the interaction between fully acculturated and embodied agents and an alien and remote world. For practical purposes, then, *string theory* as a term unifies these disparate versions into an imaginative whole. It supplies a basic container, so to speak, for theorists and lay audiences alike to amass a collection of images, ideas, and concepts according to varying criteria of inclusion. This imaginary, thus populated, affords its participants a space of possibility and interaction. String theory as a theory in the broadest sense thus functions as a placeholder that conceals a complex, arguably incoherent, multiplicity.

The problem of string theory's coherence is further complicated by the organization of high energy physics itself. In her ethnography *Beamtimes and Lifetimes*, Sharon Traweek observes that the majority of string theorists must undergo upwards of fifteen years of specialized training before they earn recognition from the community as independent professionals (74). While certain materials comprise what one could consider a pedagogical canon, the field is broad enough for different schools to emphasize certain aspects over others. The field is, in fact, so vast that string theorists are obliged to specialize even within string theory itself, at the expense of claiming authority in other subjects within the field. The number of years that a string theorist-in-training must invest before he or she is "up to speed," increases with each passing year. Currently, many must commit to an additional few years of postdoctoral research in order to achieve professional independence.[23]

It is also difficult to delineate precisely the boundaries between string theory—both its imaginative content and its praxis—and the rest of high energy physics. Much interesting and important work in string theory takes place at the boundaries of the discipline. M-theory represents one example of this, involving, as it does, supergravity theory. Some theorists, such as Michio Kaku, have attempted to synthesize string theory with quantum field theory into what is known as *string field theory*. There is also an active field of inquiry that concerns itself with a holographic correspondence between quantum field theory and string theory.[24] The following are also hybrid fields: string theory cosmogony; string theory and mathematical physics, whose practitioners, Traweek observes, "concentrate on linking developments in mathematics with ideas emerging in particle physics"; string theory and phenomenology, done by specialists "who work at finding the best fit between data and existing theories [and who] may occasionally suggest experiments" (111); and whatever recent syntheses string theorists might undertake with rival theories, such as loop quantum gravity or twistor theory. Some theorists are keen to distinguish between two camps within the discipline: model builders and theorists. Lisa Randall defines model building as "what physicists call the search for theories that might underlie current observations" (8). Model building and theorizing represent different strategies for reaching a similar goal: model builders tend to work from observational data toward a formal structure whereas theorists tend to work from the formal to the empirical. But the distinction between the two approaches would seem to be more a matter of emphasis than a categorical difference. Both approaches accommodate a multitude of methodologies.

Sociologists Martina Merz and Karin Knorr Cetina have observed that even the mathematics employed in string theory betrays a certain hodgepodge quality.

> Our [study] yielded layers of methodical policies, "ansätze," tricks and other devices, which are piled into doing a theoretical computation.

> The policies, ansätze, tricks and devices were mutually embedded in one another within a sequential interactional system involving disembodied objects, several physicists and competing teams. (74)[25]

Again, this speaks to the radical heterogeneity of string theory as professional practice, echoed here by Merz and Knorr Cetina in their description of "tricks and other devices" being "piled into" a "computation," where they become "mutually embedded in one another within a sequential interactional system involving disembodied objects" and the agents acting upon them, the "physicists and competing teams." Merz and Knorr Cetina, of course, are associating the mathematics with social practice. But if the mathematics, the "tricks and other devices" to which they refer, also correspond to images within the exposition of string theory professional practice, whether technical articles or more informal communication, then any imaginary that emerges from these texts would also exhibit a comparable structure. It would appear as an "interactional system involving disembodied objects." To speak of the hodgepodge nature of performing string theory mathematical operations is then also to implicate its concomitant imaginary.

So, in defending string theory's coherence as one complete, contiguous whole, one is forced to confront a myriad of multiplicities on various fronts—historical, sociological, formal, and textual. String theorists themselves would be the first to admit that what the theory most acutely lacks is a unifying principle of the kind that Einstein's theory of general relativity has —an example often cited when discussing this problem afflicting string theory.[26] A handful of intricately related equations —the Einstein field equations —codify general relativity into a coherent whole, which can be articulated in "plain language" as a basic, readily intuited principle. This is the famous *equivalence principle*, which states that acceleration and the gravitational force are one and the same. Traditionally reductionist physicists, such as Bohr, Feynman, and Einstein, have promoted such easily intuited principles as the litmus test *par excellence* for any theory's legitimacy. String theory lacks its own equivalence principle; nor does it possess a concise and universal canon of equations, as general relativity does.[27] General relativity represents, in some respects, the apex of triumphant reductionism. Even so, general relativity contains its own imaginary—a complex of images that correspond to the theory's mathematical expression, and which function in an expository, rather than a quantitative, context.[28] What this suggests is simply that the imaginary that coincides with a given theory—whether general relativity or string theory—will, to a certain extent, reflect the relative complexities and ambiguities of that theory's formalisms. If the real business of theoretical practice consists of manipulating mathematics, then the remainder, that part of the text that is not mathematics, is more appropriately understood in terms of an imaginary.

As a branch of reductionist high energy physics, string theory, even while assuming its own unitary "endo-epistemology," as Serres puts it, has a more difficult job in claiming scientific authority due to its all the more obvious multiplicity. In an examination of the textual, one is left to consider the extent to which an imaginary works to shore up this seeming lack of cohesion. Furthermore, it would be paramount to explore the extent to which, in a given discourse, its imaginary draws from and relies upon an imaginative *doxa*—a conventionalized epistemological imaginary that precedes and makes possible the ostensive novelty of string theory images. Paradoxically, it is this reliance on the part of string theory—its embeddedness within—an enveloping epistemological imaginary, that allows for the possibility of its reception by a given audience as comprehensible, coherent, and total. This, in turn, prompts the question: if it is highly problematic to speak of even an individual string theory imaginary as less than hopelessly heterogeneous, then what are the cultural conditions under which both the professional community and a lay audience have come to understand string theory as string theory and not as a cacophonous medley of hadronic, bosonic, and supersymmetric string theories, of brane theories, M-theories, F-theories, and Landscape theories, potentially *ad infinitum*?

A Return to the Eleventh Dimension

To conclude this chapter, I return to the disquisition offered up by Scott Mehring of Mechanicsburg, Pennsylvania. His narrative invokes the following images in succession: the Big Bang, elements, atoms, neutrons, protons, stuff, spontaneous generation, a theory, meat, the ground, maggots, physics, the eleventh dimension, infinite universes, death, the Earth, the Bible, everlasting life, an intelligent designer, complication, one dude, the very top, everything, gods, goose bumps, an endless loop, computer programming, and insanity. Of these, the core images would seem to be: time, elements, life, death, God, infinity and eternity, and the eleventh dimension. As mentioned earlier, the eleventh dimension is an image adapted specifically from string theory. Its inclusion, along with the scientific ideas of atoms, neutrons, protons, the "dead theory" of spontaneous generation, and the Big Bang, mark the imaginary as scientific. Yet the imaginary also seems to assume that science progresses through history and that older theories are invalidated and replaced by newer ones.

The human agents who operate within the imaginary are "you," "me," and "they." They act within the imaginary through the following verbs: going back, thinking, saying, believing, getting, hopping, going insane. Both the images and the actions related to them exhibit a wide range of qualities—some are highly abstract, some concrete, others blends of the two. If one visualizes the cosmos that Mehring has articulated here, it seems to consist of several layers, where (an ostensibly disembodied or ethereal) thinking you and I surround and penetrate in toward an interior. This is followed by several creating and anthropomorphic gods ("one dude") the eleventh dimension, elements that generate and then constitute an infinitude of universes, various times corresponding to those universes that all begin with their own Big Bangs, the Earth with a ground where maggots and human bodies are born and die, while human souls, upon the death of the body, "hop" from universe to universe within the eleventh dimension. Mehring conflates the elements he evokes—the neutron, the proton neutron, stuff—with God, as a creator or intelligent designer. This cosmos, furthermore, has an overarching structure of an endless loop—presumably implying that the infinities stretch out to such an extent that they eventually return back to their origins. And through excessive contemplation of this vast spectacle, the thinking you and I, both as disembodied souls within the narrative, and as persons outside it looking in, risk insanity. By insanity, Mehring presumably means a kind of paralysis whereby "we" become incapable of either thinking or acting in any meaningful way. There is a schematic tension here between mobility—both in mind and body—and paralysis.

It would be easy to dismiss this imaginary merely as hopelessly muddled—the bemused ramblings of a provincial crackpot, as the journalist Chapman undoubtedly expects the reader to do. But even this initial inventory reveals a certain noteworthy complexity and intricacy to it; a rich space of meaning. The principal agents within the imaginary—the "you" and "I"—face a universal quandary; namely, the problem of death. We are, in turn, offered an eschatology, a meditation on a possible escape from death, in the form of a transmigration. On the one hand, we are able to escape the bondage of decay and death (alluded to by the proximity in the narrative of meat on the ground and the maggots that consider that meat as food). On the other hand, contemplation of the infinite regression of universes and creators inherent in this "new theory, the eleventh dimension" jeopardizes both the contemplator's mental health and his imaginative motility. In this, Mehring seems to recognize that to embrace theoretical physics requires belief, that there is no preordained correlation between theory and truth.

In Mehring's imaginary, the eleventh dimension is a container: it contains an infinite number of universes.[29] He also posits the "we" as dual—a Cartesian (or Christian, for that matter) *res cogitans* and *res extensa*, a soul that inhabits the body like a homunculus, but upon death escapes. I say homunculus because this I-soul or you-soul continues to exhibit embodied characteristics: it can "hop from universe to universe to universe." The schema that accompanies this image—of a soul in the form of a body—brings about a radical transformation in the scale of these universes. What one would normally consider unimaginably vast becomes condensed to the scale where, in the analogy, the soul-body is able to hop—generally imagined as a rather small jump—from one universe to another. Perhaps Mehring imagines hopping to be virtual, more in keeping with a *Star Trek*-style teleportation where the body disappears from one universe and magically reappears in another. But he does not make this explicit. Later, one is confronted with yet another radical transformation of scale: this whole imaginative structure gets compacted into a "loop." Mehring goes on to connect this "new theory" with Biblical eschatology—with the Christian promise of an "everlasting life." In a leap of logic, he then problematizes another salient assertion of Christian dogma —often called the cosmological or 'Prime Mover' argument —that *one* God created the universe.

Mehring's imaginary is essentially inclusive; through it he tries to synthesize various seemingly disparate knowledges —string theory, Big Bang theory, and Biblical theology—into a coherent whole. Obviously, one could debate the seriousness or success of the effort. But it is undeniable that the imaginary has a fairly intricate structure—drawing on a wealth of images that bear with them various schemas, some readily compatible, others not—and that the synthesis of these images demand a certain balance between mobility and stability. The imaginary suggests both freedoms and constraints, possibilities and limitations. But what is most interesting about Mehring's adaptation of a string theory imaginary is how its image of the cosmos couples with or reflects Mehring's own presumed social identity. His cosmology represents, in a certain sense, either a prescription on how to live one's life—or a justification for a life already lived. In analyzing Mehring's imaginary, one discerns in it the following preoccupations: a fear of death and longing for escape from it; a desire for absolute freedom of movement, for the abolition of all restraints on the liberty to gratify one's impulses; a close association between mental activity and visceral bodily responses, between contemplation and goose bumps; and a fear of loss of control implicit in the final caveat concerning insanity.

As was pointed out earlier, Mehring's disquisition arrives pre-framed by a judgment implicit in the details Chapman chooses to disclose to the reader. These details classify Mehring as a socially positioned agent and tacitly grant the reader permission to assume a critical, if not ironic, distance from the imaginary he presents to us in his speech. This framing makes Mehring's imaginary seem hopelessly incoherent, logically suspect, frivolous, and utterly pseudoscientific. In addition to the merit of a scientific imaginary's content—with which readers must necessarily have some degree of familiarity in order to assess—our predisposition to judge it meaningful is predicated on our submission to the authority of the proponent of that imaginary. A scientific imaginary such as that of string theory is framed and supported by institutions of belief that are tied to a particular epistemological imaginary —a way of knowing the world in its totality as a given order, an ordered whole. As such, an epistemological imaginary functions in two dimensions. On the one hand, it works, from the outside, to authorize the value of the scientific imaginary to be consumed, and on the other hand, it arrives embedded within the imaginary in the tight coupling of social agencies with a presumed cosmic order. This coupling constitutes a double-bind: social institutions outside the scientific imaginary sustain it, while the imaginary, remote from daily social experience, delivers within its own structure representations of social practice consistent with the institutions that reinforce it.

Decoding the culture implicit in Mehring's imaginary, one finds, then, a soul—disembodied yet possessing abstracted bodily features, isolated from its brethren, "hopping" somewhat indiscriminately from one gratification to the next, without much sense of purpose or even social obligation. A culture of individualism provides inspiration for Mehring's imaginary, while the imaginary, in turn, justifies a certain solipsism.

CHAPTER 3

The Romance of Encounter

String Theory Technical Discourse

These ambiguities, redundancies, and deficiencies remind us of those which doctor Franz Kuhn attributes to a certain Chinese encyclopedia entitled *The Celestial Emporium of Benevolent Knowledge*. In its remote pages it is written that the animals are divided into

1. belonging to the Emperor
2. embalmed
3. trained
4. piglets
5. sirens
6. fabulous
7. stray dogs
8. included in this classification
9. trembling like crazy
10. innumerable
11. drawn with a very fine camelhair brush
12. et cetera
13. just broke the vase
14. from a distance look like flies

(Borges 103)

Exposing Technical Exposition

In his preface to *The Order of Things*, Michel Foucault writes of how reading this passage from Borges caused him to laugh out loud; that it, as Foucault puts it, "shattered . . . all the familiar landmarks of thought . . . breaking up all the ordered surfaces and all the planes with which we are accustomed to tame the wild profusion of existing things" (xvi). What Foucault so appreciates in Borges's taxonomy is its vivid demonstration that how we categorize the things of the world in large part determines our sense of order in the world. But just as those categories are by no means inherent to the world, any taxonomy our categorization generates can only offer a provisional, and more often than not, internally contradictory coherence. A close scrutiny inevitably will yield certain "ambiguities, redundancies, and deficiencies," especially when that perspective is remote from the tradition that institutionalizes a given taxonomy—or, in other words, when that taxonomy does not embody common-sense. Borges stunningly illustrates the assumption that the act of categorization is first and foremost an act of imagination—that it is an imaginative achievement. And that imaginative achievement becomes all the more extraordinary if one is willing to acknowledge that much more is at stake in the act of categorization than, as the classical theory would have it, the mere grouping of things based on the shared properties of that group's members. Pivotal to many of our conceptual categories are images, which serve as prototypes for that amalgam of ideas and experiences that have come to define a particular phenomenon as, categorically, that which it is, according to our understanding, even when that knowledge is scientific.

In some respects, what string theory offers us is an imaginary akin to Borges's "Celestial Emporium of Benevolent Knowledge": it is a radical recategorization of basic things in the world. But much of what is so perplexing about these new kinds of things on whose existence string theory makes its mathematically authoritative claims, is that they—strings—are abstract phenomena that undergo abstract transformations within abstract and extraordinarily remote (microscopic) spaces, and yet they nevertheless have a "physical interpretation." This chapter concerns itself with a selection of string theory technical articles in their presentations of just such abstract spaces. The central question is: precisely what is at stake in a given technical article when a theorist ventures a "physical interpretation" of that abstract space and the phenomena that populate it? Or in other words, what is at stake when a technical article exposes—by means of exposition—string theory as an imagined space?

In chapter 2, as a variation on Le Doeuff's philosophical imaginary, I defined a scientific imaginary as a complex of images set off from but problematizing concept. In a scientific enterprise such as string theory, these images serve as doubles for mathematical arguments. Yet the images possess their own inferential relations distinct from this structure of correspondences. A congeries of images that in one valence are bound individually in one-to-one correspondence to mathematical expressions, in another valence, together constitute an autonomous whole, an imaginary. Such an imaginary is marked as scientific inasmuch as it retains its reference to what a culture recognizes as scientific practice—in the case of string theory, with a particular, dynamically defined amalgam of methods and reasons, empiricisms and logics, with an emphasis on the plural. In general, string theory technical articles consist of two basic components: mathematical arguments framed by expository prose. By the theorists' own admission, the mathematical expression (propositions, calculations, proofs, ansätze, etc.) embodies the core content of a given article. It is in the mathematics that theorists constitute the solutions to the problems that they consider meaningful. In this account, exposition, as handmaiden to the mistress mathematics, helps to orient the reader—to offer didactic and analogic guidance. And yet the mathematics must ultimately stand on its own merits. Accordingly, a technical article's exposition represents, in some respects, the reformulation of a discourse that does not require reformulation. Paradoxically, the doubling images make the imaginary to which they belong redundant.

It is useful to make a distinction, however provisional, between technical terminology that serves as markers for mathematical expressions within the technical articles considered and physical interpretations of those expressions. Much of what a non-specialist would recognize as impenetrable jargon within the exposition of these articles functions in a relatively straightforward way: it is meant to help orient the reader through a thicket of dense mathematical apparatus. While a neat distinction between such technical jargon and terms that are meant to evoke physical phenomena is not unproblematic, for the purposes of this argument, let us envision a procedure of differentiation that produces a spectrum, where one pole represents terms that designate highly specific mathematical expressions (for example, "Majorana-Weyl light-cone spinor," "Dirac brackets," or "Fock space states"), while the opposite pole coterminates with what the theorists would consider, in their implicit epistemology, a physical phenomenon; for example, the string or brane. The close reading that follows will focus on the latter pole in an effort to determine the extent to which the inherent heterogeneity of the images at stake, their "ambiguities, redundancies, and deficiencies," inevitably compromise coherence of any concomitant imaginary.

The abstract theoretical spaces within string theory technical exposition are macaronic in large part due to the imaginative resources available to the theorists in their efforts to engage actively with such spaces. As such, the following analysis does not assume an unambiguous relationship between a given term and the image that it may or may not evoke.

The correspondence between term and image may undergo slippage on account of various factors, including, but not limited to: the unique associative proclivities of an individual, conventions based on a given culture's linguistic tradition, and the inherent ambiguities between language and non-linguistic cognitive capacities. These respective factors further contribute to incoherence. Nevertheless, in certain cases, it is worth exploring the etymologies of key theoretical terms in order to better understand any potential ambiguities or associative resonances latent in those words. Within abstraction, we find not so much a vestigial concretion that plays a substantiating role in a given imaginary, but rather, an imaginative virtuality predicated on the relation of the images at stake to the imaginative complex itself. In effect, concretion is something imaginatively mediated and context-dependent.

Before we consider strings and branes specifically, it is also worth briefly examining how physicists in general name the theoretical objects with which they work. To my knowledge, there exists no universal prescription for naming conventions within theoretical physics. Theorists tend to name things, whether objects or operations, according to one of the following four rationales: names that are what a consensus would deem commonsensical, names that are whimsical, names that serve as a form of credit dispensation, or as a form of titular homage. Take, as examples of a commonsense rationale, the particles that appeared in early versions of string theory: partons, mesons, and hadrons.[1] Richard Feynman presumably named the objects he theorized, *partons*, by lopping the *-icle* off of "particle" and adding the suffix *-on*—a convention that quantum theorists had long ago adopted to designate certain kinds of subatomic particles.[2] *Parton* seems to be a commonsense appellative that, however generic sounding, does not bear along with it an excess of etymological baggage—simply that which is, in the appropriate context, part of a whole. Mesons were initially called *mesotrons* by theorist Hideki Yukawa in 1935: a portmanteau of the Greek *meso-* for middle, with the *tr* from electron and neutron, along with the suffix *-on.* A few years later, this name was corrected by Werner Heisenberg to more accurately reflect its proper Greek etymology.[3] With a similar rationale, the term *hadron* comes from the Greek root meaning "thick, bulky" with, once again, *-on* added.

An example of the second rationale, a term based on a particular theorist's taste, is *quark*. As mentioned in the first chapter, Gell-Mann professed to have named the quark after a passage in James Joyce's *Finnegans Wake:* "Three quarks for Muster Mark!" (383). In Gell-Mann's theory, the quarks that constitute protons and neutrons come in sets of three. A whimsical name like quark perhaps exacerbates the oft-marveled-at strangeness of the microscopic quantum world. The *Veneziano amplitude* is named thus, in keeping with the third rationale mentioned earlier, in order to give credit to the inventor. An example of the fourth rationale, titular homage, are the classes of particles known as *fermions* and *bosons*, respectively. Paul Dirac named fermions in honor of the Italian physicist Enrico Fermi; he named bosons for the Indian physicist Satyendra Nath Bose. Both had worked on problems that Dirac's solutions brought to a satisfactory conclusion. Within contemporary braneworld modeling, *Kaluza-Klein particles* are named in deference to the early twentieth-century contributions of those two theorists in their pioneering attempt to reconcile general relativity and quantum electrodynamics.[4]

The cumulative impression of this ad hoc collection of subatomic particle names in quantum theory seems to resist easy association with more familiar objects or experiences. They almost feel calculatedly weird; compounded, in the case of the quark, by its subcategorization into six types: *up* and *down*, *strange* and *charmed*, *top* and *bottom* (also sometimes called *truth* and *beauty*). In keeping with the erratic pace and piecemeal nature of progress in theoretical physics itself, theorists do not seem to be systematic about the names they give theoretical objects, opting, at any given christening opportunity, for a name based on one of the four rationales. At times, a name may bear with it connotations that make sense in the context of the mathematical argument to which it refers. Yet, at other times, such a correspondence between image and concept can only ever be motivated, but not self-evident. By motivated, I mean the rationale for a given term's selection may be retrospectively gleaned and may make sense—yet its appropriateness is neither completely arbitrary nor predictable. Furthermore, certain contradictions, paradoxes, or counterintuitive inferences inevitably reveal themselves under scrutiny. The presumption seems to be that since names only serve as placeholders for the more precise mathematical expressions to which they correspond, not much care need be given to what motivates their fabrication. All the seriousness and precision goes to concept, leaving the names—and the images they conjure—at complete liberty to constitute a surplus of meaning, depending on the context and the rationale employed at their origination. As such, the images chosen to double for mathematics all possess a certain cultural specificity, if not outright arbitrariness (as in, for instance, the term *quark*). Ultimately, the arbitrariness of such imagery would then have wider, perhaps inadvertent, implications for the imaginary that they comprise. By not taking an imaginary seriously, string theorists are then free to displace certain "unspoken work" onto it—"the work," as Le Doeuff puts it, of "dogmatization." This work of dogmatization is twofold: it is the displacement of incoherence away from mathematical argument into an imagined space and, concomitantly, the epistemological facilitation of an imagined engagement with that space. Such an abstract theoretical space provides for the possibility of not only engagement but, importantly, encounter between the theorists and the phenomena that occupy that space. This imagined encounter necessarily incorporates an abstracted form of agency on the part of the theorists—and, additionally, the phenomena duly encountered.

For the purposes of this chapter, I have chosen three technical articles to read closely. They each represent an imaginary at an important moment in string theory's historical development. The first is by Leonard Susskind, titled "Dual-Symmetric Theory of Hadrons—I," and published in the journal *Nuovo Cimento* in 1970. As mentioned in the introduction, this article represents the moment in the history of string theory when the term *string* first appears in the technical discourse—within a model that seeks to formalize the heretofore elusive dynamics of the atomic nucleus—and is recognized as something theoretically viable. Susskind's paper establishes the basic imaginary for hadronic string theory—string theory's formative early incarnation. The second article represents superstring theory at the most sanguine zenith of the "first superstring revolution" of the mid-1980s. Authored by David Gross, Jeffrey Harvey, Emil Martinec, and Ryan Rohm, the paper, titled, "Heterotic String Theory (I): The Free Heterotic String," and published in *Nuclear Physics* in 1985, proposes a particular formulation of superstring theory that many theorists consider—with some supplemental modifications—to be the most realistic of the five versions developed at the time. The third article, by Lisa Randall and Raman Sundrum, "An Alternative to Compactification," was published in *Physics Review Letters* in 1999. An important and highly influential paper within the field, it typifies the kinds of modeling being done in the period following the "second superstring revolution." In keeping with the strategic turn of M-theory, the paper deploys branes as its fundamental object in a novel way, and by doing so, challenges some of the basic assumptions of superstring theory and supergravity before their suggested synthesis by M-theory.

These three texts represent a snapshot survey of major moments in string theory's development with novel formulations of the string theory imaginary: the hadronic string as the strong force that binds quarks together in the nuclei of atoms, the heterotic superstring as the basis of a fully supersymmetric particle spectrum—that includes gravity—in ten space-time dimensions, and the three spatial dimensions of our universe as a "3-brane embedded in five dimensions" where "Newtonian and general relativistic gravity is [*sic*] reproduced to more than adequate precision" (Randall and Sundrum 1). Two important concerns impinge on "physical interpretations" of mathematical arguments coupled to imagery manifest in exposition: firstly, the extent to which an imaginary—as an abstract space—constitutes a coherent cosmic order, and secondly, the ways in which an imaginary provides for imagined encounters between an abstracted human agency and what the theorists tacitly mark as natural phenomena.

Before tackling these articles, let us briefly return to the binary of concrete and abstract, discussed in chapter 2, and of particular significance to string theory technical exposition. *Concrete* literally means "that which has grown together." The commonsense definition of the qualifier concrete is that it describes objects both solid and close to hand; things that have a direct and tangible impact on our embodied experience. In contrast, the abstract is that which has been drawn out or away—higher-order phenomena that are immaterial and difficult to grasp directly. Yet with respect to string theory, that which is concrete or abstract cannot be disentangled so readily. On the one hand, string theory pertains to an abstract space—one whose truth-value rests with formal arguments that describe the dynamics of theoretical objects on scales remote from perceptual categories relevant to the human body. All access to strings must be mediated via mathematics and (possibly) instruments. We are removed from the string in its "natural" domain by multiple layers of abstraction, by multiple iterations of a drawing out. On the other hand, theorists argue for the reality of the objects they describe—strings and branes.[5] They are objects gathered together in the sense that they exhibit a certain cohesion and redundancy—their form persists, however dynamically, through time. As such, they may be brought close to hand: though mediated via machines, we may be able to not only "observe" them in nature but "harness" them as well. From a realist position, the realm of the string, though an abstract theoretical space, is also concrete inasmuch as it eventually may be observable and manipulable by means of probes; for example, the Large Hadron Collider or some future supercollider.

Mathematician Robert Osserman argues that, with respect to physics, abstraction functions essentially in three ways. Firstly, it affords what he calls the "power of universality, allowing a single rule to apply in very different circumstances" (145). Secondly, mathematical abstraction "brings clarity to what may be a confused situation" (145). And lastly, abstraction "provides us with great freedom to let our imaginations roam, permitting us to devise new and alternative versions of reality—versions that may or may not correspond to something in the real world" (145). Osserman's definition of abstraction is persuasive in that it reflects our commonsense notions of its contrast to the concrete. Yet these three functions of abstraction may apply equally well to that which we are inclined to consider the very essence of concrete; for example, an apple. If we make a distinction between a particular physical instance of an apple—of which we enjoy only a partial and cognitively mediated access,[6] and the idea of an apple by means of which we recognize all apples as "an apple," then it becomes apparent that the apple, heretofore unambiguously deemed concrete, is also abstract. The image of an apple is universal: it allows one to recognize and reduce a disparate multitude of "different circumstances" as instances of the one prototype. Furthermore, the idea of the apple brings "clarity" to a "confused" situation. In learning to recognize all apples as instances of "an apple," a profusion of phenomena has been tamed, allowing us to function ever more effectively in the world. And lastly, the idea of the apple equally allows our imaginations to "roam" in that we may endlessly tinker with the possible permutations of the prototypical apple such that it does or does not retain its essential apple-ness.

Commonsense definitions of the concrete that rely on conditions of human-scale proximity and graspability break down when we acknowledge that the natural kinds of objects that we take for granted do not necessarily exist *a priori* out in the world. While there certainly seem to be patterns and regularities in the "objective" world independent of the intervening mediations of human embodied cognition, the way those patterns and regularities appear to us in our active (albeit mostly unconscious) taxonomic apprehensions, in many respects, constitute such natural kinds both as natural and as kinds. Accordingly, that a Fock space is more abstract than an apple—or a superstring, for that matter—is more a convention predicated on the feedback delays between body and cognitive processes of ideation conditioned by culture. With an apple, these two events are closely coupled—the experience of interacting with an apple and recognizing such an experience as involving an apple. No phenomena arrive before our bodies as unstructured "naked" experience; cognition is such that experience comes already prestructured.[7] With string theory, unlike the ostensive immediate proximity of the apple, the delay yawns considerably: complex, time-lapsed, multi-tiered interactions with mathematical constructions and machines reinforce the notion, initially abstract, that the string is indeed, on the contrary, concrete, and thus real. Image and experience must be brought together only after a meticulous and ongoing process of mathematics- and machine-mediated interweaving. While all spaces are abstract in the sense that they are imagined spaces, the spaces presented by a string theory imaginary only earn concrete status after numerous cycles of the forging of communal consensus and of prosthetic intervention.[8] Given the entanglements of concrete and abstract, how then can we more precisely define the space of string theory technical exposition?

I do not intend here to offer an exhaustive analysis of the concept of space, but rather, to highlight certain specific images of space that are particularly pertinent to its role within a string theory imaginary. More fundamental images of space emphasize the notion of place—the ground upon which a body stands. In ancient Greek, space is synonymous with *topos*, or place. In Latin, *spatium* comes from the verb *spatiārī*—to walk, extend. The body defines space by its movement over ground, from one place to another, where space is understood as an aggregate of many places. The rudiments of the further abstraction of space are made explicit in a text such as *De Rerum Natura*. There, Lucretius delineates nature thus: "All nature then, as it exists, by itself, is founded on two things: there are bodies and there is a void in which these bodies are placed and through which they move about" (420). Lucretius's image of space abstracts the hitherto intuitive relation of body and place. Space is that which frames and distinguishes objects in the world, where space (the void) is the ground and objects are situated as figures within that ground. The contrast between a body and the space it occupies is necessarily more subtle in Lucretius's formulation. The body does not merely stand on a place and move from one place to another; rather, space is defined by the gaps between bodies. Space is the absented presence of (mobile) bodies. But this formulation prompts the question: do bodies then displace space or does space (the void) permeate those bodies as well? This may seem like an antiquated, if not trivially metaphysical question, but it echoes, in some respects, the epistemological ambiguities that arise in a contemporary string theory imaginary.

In *Genesis*, Michel Serres defines space in a way that further abstracts this relatively literal notion of it as an aggregate of the places our bodies have tread over and back again. "Space is what we call a relatively homogeneous, isotropic multiplicity, subject to some law or definition. It is always necessary . . . to have a certain redundancy in order for a space to be, in order for a space to be thinkable" (116). For Serres, space is essentialized as redundancy. Where there is no redundancy, there is no thinkable space. Serres's definition of space presupposes redundancy as a necessary condition for the existence of not only space, but cognition as well. Space manifests through its repeated recognized occupation. It is this act of repeated recognition of space that allows space to exist. Since one fundamental structure, as a necessary condition, of such space is redundancy, a space visited with some frequency would become a familiar space. It is in this sense that the redundancy implicit in repeated visits, both physically and cognitively, leads to the apprehension of a space as homogenous, as "subject to some law"—a law bound up in our experience of that space as familiar. The very notion of space thus implies a certain domestication.

I offer one more image of space to further highlight the extreme motility of its significance—of the epistemological slippage in our images of space between the concrete and the abstract. The physicist Lee Smolin claims that: "there is no meaning to space that is independent of the relationships among real things in the world. . . . Space, then, is something like a sentence. It is absurd to talk of a sentence with no words in it" (*Three* 18). In Smolin's formulation, an alternative abstraction reverses the figure and ground of the Lucretian imaginary. Space is not something that objects occupy, that, as a consequence, has the effect of distinguishing one object from another, but rather, is wholly constituted by the relation of those objects to each other. In this conceptualization's extreme form (which, granted, would be an interpretive departure from Smolin), the "real things" themselves become nothing more than aspects of their interactions and it is those interactions that constitute space—phenomena as aspects of space. The analogy would then be that words have no meaning, save within the context of the sentence.

By incorporating time, we are now in the position to imagine these contrasting images of space as events, where one version privileges the object and the other, location. In *Philosophy in the Flesh*, Lakoff and Johnson argue that "we do not perceive [events] that are neutral between figure and ground. . . . Perception requires a figure-ground choice," what they call "event-structure duality" (198). In an object-event, "causation is the transfer of a possessible object (the effect) to or from an entity (the affected entity)" (199). In a location-event, "causation is the forced movement of an entity (the affected entity) to a new location (the effect)" (199). Lakoff and Johnson write:

> In both cases, the figure is conceptualized as moving and the ground as stationary. And in both cases, there is a causal force that is applied to the figure, moving it with respect to the ground. But in the two cases, the causal force is applied to different things. In the Location case, the causal force is applied to the affected party, since it is the figure. In the Object case, the causal force is applied to the effect, since it is figure. (199)[9]

As we shall see, the technical articles considered in this chapter repeatedly make use of these contrasting image schemas of events, where the figure and ground of object and location are shifted to suit the demands of the imaginary. In some instances, counterintuitively, it is the abstract theoretical space that is the figure, and the objects that populate that space are imagined as the ground.

An abstract space is first and foremost a space that is imagined as being far off. Yet paradoxically, an abstract space, through the very process of abstraction, is made proximal within an epistemological imaginary. It is that which has been drawn from far off but brought ever nearer such that the imaginary engenders some tentative modicum of redundancy and familiarity. To return to Osserman's observations on the utility of abstraction, the power to be able to apply a single rule to a multitude of circumstances affords the wielder of such a power a certain comfort and satisfaction. The confusion of the multitude, its chaotic indecipherability, becomes tamed. This taming of the chaos of the indeterminate, this determination by abstraction, in turn, allows theorists to imagine themselves moving comfortably and decisively through this freshly demarked space.[10] It grants the imagination mobility, for the imagination cannot move freely within a thicket of confusion. It tends to freeze, become rooted, excluded, hesitant, muddled. In this sense, the close reading that follows explores the extent to which an abstract space enables quasi-domestication. An imagined space typifies a process of quasi-domestication in that it continually betrays its status as a space that is still far off yet being drawn ever closer. Such an imagined space both invites and resists domestication. It is both sensible and insensate, intuitive and counterintuitive. It draws on familiar images and concepts, hints at the concrete, that which is close to home, while also throwing up barriers to such intimacy. Accordingly, the further off such an imagined space feels, the more counterintuitive its formulation manifests—literally, such an imagined space and the objects that populate it are against any seeing in.

The Hadronic String

As mentioned earlier, Leonard Susskind is the first theorist to use the term *string* to describe the behavior of hadrons, the class of heavy particles that make up the nucleus of the atom.[11] But before delving into that particular technical article, some background context is warranted. Most string theorists recognize the publication of Veneziano's 1968 paper[12] on the scattering amplitudes of hadrons as a pivotal moment for the inception of string theory. The principle object of analysis in the article is pion-pion interactions, pions being a type of hadron. Veneziano argues that:

> in the scheme proposed here each trajectory is not really an independent object. In most cases (as in $\pi\pi$ scattering) more than one trajectory *must* coexist in a kind of «conspiratorial» situation, namely with certain relations among their trajectory and residue functions. (196, emphasis in original)[13]

As the two pions collide and a residual scattering of by-products occurs, certain trajectories only manifest themselves in strict relation with other trajectories—what he calls a "conspiratorial" situation. In this interpretation of the interaction, pions and pion byproducts remain fundamental objects, whereas the trajectories that mark their interaction exist only inasmuch as they must mutually inform one another in the literal sense of inform—namely, that interaction generates form. These trajectories are short lived—the indirect traces, recorded by detectors, of the violent collision that has just occurred on subatomic scales. In Veneziano's imaginary, pions have substance, but trajectories do not, an image all the more emphasized simply by calling them trajectories—a path through time, an abstraction, ephemeral and insubstantial. This is a distinction that takes advantage of the ambiguities in a theoretical physics imaginary between notions of matter and energy, a distinction largely dependent on a precise definition within a specific context.[14]

At play in Veneziano's "scheme" of pion-pion scattering are two contradictory image schemas analogous to the abstractions of space exemplified by Lucretius and Smolin, which are predicated on two basic images that reverse the relative positions of figure and ground. Since pions are imagined as objects in Veneziano's scheme, their interactions (and subsequent "scattering") cannot be, by the logic of the image-schema, "independent object[s]." These trajectories, as "relations," are thus imagined as insubstantial—events with "residue functions." As we shall see, this fundamental paradox between object-event and location-event image schemas will play a pivotal role in the three technical articles that follow.

In 1969, Leonard Susskind published a short paper in *Physical Review Letters* titled "Harmonic-Oscillator Analogy for the Veneziano Model." He begins the article: "We present a model scattering matrix based on a relativistic harmonic oscillator" (545). In Susskind's imaginary, Veneziano's mutually informing trajectories become a harmonic oscillator, a theoretical object in its own right with the key attribute that it vibrates.[15] A year later, Susskind proposed a system that elaborates on this new understanding of hadrons as harmonic oscillators in his paper "Dual-Symmetric Theory of Hadrons—I": "In what follows we shall consider this system to be a model for the internal structure of a meson" (463).[16] As we have seen, quantum theory defines quanta, its fundamental objects, principally in one of two ways, depending on the context. For instance, a meson can be imagined either as a point-particle or as a wavefunction. The wavefunction represents a probability distribution that bears information concerning the meson's relevant attributes—for instance, its location in space-time, its momentum, its spin and charge. Susskind's system adapts Gell-Mann's theory of quarks, which argues that mesons are not simply point-particles, but actually possess internal structure. This image of internal structure overwrites the image of the meson as point, which, by definition, cannot have any structure inside of it, since it lacks spatial extension. It is utterly counterintuitive to imagine a point having internal structure: in order to cohere, the one image must supplant the other. In Susskind's revised imaginary, a meson becomes container-like with internal spatial extension that can thus possess structure. In fact, Susskind's imaginary has become more heterogeneous in that it now must accommodate the contradictory images of particle-as-point and particle-as-container with internal structure.

Within the imaginary, this supersession of point by container prompts further explication. Susskind describes the internal structure of this container-like object as "the mechanics of a harmonic system composed of an infinite number of mass points, each coupled to its nearest neighbor by harmonic oscillator potentials" where "the system must be four-dimensional instead of three-dimensional" (462). Now one must imagine, within the container that is the meson, an "infinite number" of points "coupled" to adjacent points by "potentials." These "mass-points," as Susskind calls them, coupled as they are, provide for motion such that the entire mechanism oscillates. Yet one must imagine that this mechanism occupies "four-dimensional" space. The container must have hidden extra room, so to speak, since intuitively, a normal container only has three dimensions—length, depth, and breadth.[17]

The counterintuition of four spatial dimensions is compounded by the image of an "infinite number" of points. Analyzing the concept of infinity as an image, Lakoff and Rafael Núñez suggest that "processes that go on indefinitely are conceptualized as having an end and ultimate result, an infinite thing" (158). This, they argue, is fundamentally an imaginative leap where the "end result" contains the gathered-together indefinite multitude—many phenomena are imagined as one phenomenon, infinity. This is akin to an asymptote (the line that some types of curves approach but never quite touch as they move toward infinity). A mathematical procedure never actually reaches infinity, only perpetually but indefinitely approaches it as an idealized arrival place; it is an implied continuation of activity. A rigorous and coherent mathematical operation corresponds here to a paradoxical image.

In order to, as Susskind puts it, "do quantum mechanics with this system, we expand" one component of the system using "creation and annihilation operators" (463).[18] After several more mathematical operations, Susskind is prepared to differentiate his system from its quantum mechanical predecessor.

> The entire space of states generated by the action of the [creation and annihilation operator] on the ground state . . . is the space of possible states of a single meson. The [ground] state is of course not the vacuum. It is the ground state of a single particle. . . . The operators . . . do not add particles to the system. Instead they excite the particle into rotational and vibrational states. (463)

On the one hand, Susskind is describing straightforwardly a series of coupled quantum harmonic oscillators where the "action" of "operators" generates a "space of states." In this sense, such terms are mathematical expressions with precise technical definitions. On the other hand, they also represent procedures that the theorist must undertake in the abstract. In this complementary sense, one may imagine this system as a mechanism that has certain components that act, once set imaginatively in motion by the theorist, and as a result of that action, they produce something by means of an involuntary and automatic agency. What these operators produce is a "space of states" that are not just any states, but more emphatically, "possible states," akin to a quantum mechanical wavefunction.

Again, in this imaginary, there is a specific and peculiar blending of object-event with location-event image schemas. Operators perform an action on a "ground state" which "generates" another space, "the space of possible states of a single meson . . . a single particle." These states, in turn, are not additional particles, but rather, "rotational and vibrational states." Objects-as-agents act on one state, the "ground" state, where "state" is understood as a configuration—a particular set of attributes of the meson. "Ground" here is understood as the system's lowest energy state. But paradoxically, one is asked to imagine an object in an abstracted location (a space) as having a set of attributes which is also a location. In addition, one must imagine a space as an indeterminately large (infinite) amassing of "possible states."

The operators act but "do not add particles to the system." In the one reference frame, these operators would appear to be inside the meson, as components of its internal structure, where they "excite the particle" to rotate and vibrate. Yet in another imaginative frame, the operators act on a ground-like state, and out of this action, a supplemental space emerges (is "generated"). The description would seem to oscillate between an object-event frame and a location-event frame, depending on the demands of the mathematical argument, the abstract "mechanism" for which the imaginary doubles. An object—as a container—possesses internal structure. This internal space evokes a counterintuitive quality—it would appear to occupy more space than the imagined boundaries of a particle as tiny container would allow: a space of innumerable (infinite) states. This imaginary is further rendered extraordinary in that these innumerable states are not actual states but "possible" states, a wavefunction, which extends into four spatial dimensions. Stacked atop this mélange of seemingly conflicting image schemas is the image of rotation and vibration. The words themselves evoke an imagined kind of motion that can be readily seen and felt. And yet there is an inherent ambiguity in what object one is meant to imagine rotating and vibrating. A fundamental particle, by definition, lacks spatial extent and thus is incapable of behavior normally attributed to objects with width, depth, and breadth; for example, a ball. If one imagines the meson as a ball, where the complex internal structure is crammed into its interior, then the notion of it spinning is coherent.[19] On the other hand, if one imagines the meson as a field-like space, or even a stacked aggregate of dynamic fields, then one is more disposed to imagine it vibrating, much in the way the water in a pool can be made to vibrate in the form of waves. Again, the reader is asked implicitly to imagine the meson in two contradictory imaginative frames of reference—as an extended object contrasted from a background space and as a specific wave-like location *within* a dynamic field-like or pool-like space.

My point in explicating the imagery in this passage is not to simply summarize an expository account of the technical discourse itself, but to highlight its macaronic quality. The internal structure of the meson upon which Susskind elaborates is not difficult to imagine because it is technical. The mathematical argument that accompanies the imagery has an internal coherence—and thus a prima facie persuasiveness—that either is or is not axiomatic to a professional reader. Whatever incoherence the concomitant exposition projects results from the inevitable shortcomings of any unitary human-scale imaginary in representing the system's intricacies. The nature of such imaginative work lies in the relative liberty the text has in mixing and matching imagery to suit the demands of exposition. In this highly heterogeneous imaginative environment it is difficult, to say the least, to distinguish between mathematical jargon and "objective" phenomena, between the linguistic artifacts of professional practice and "actual" physical objects.

Thus far in the article Susskind has only made an oblique reference to the string. In the introduction, he suggests that a "typical example" of a "simple field theory" with an "infinite number of internal degrees of freedom" is a "harmonic continuum, such as a violin string or an ideal rubber band" (458).[20] Shortly thereafter, he writes: "Let us suppose that mesons are composed of bound [quark-antiquark] pairs" (459). This, on first impression, simple image—that of two objects bound to each other—provides the core impetus for his investigation. But the image is not so coherent when one considers the nature of an antiquark—the antiparticle complement to the quark. In quantum theory, antiparticles are particles of the same mass as their complement but with an opposite charge such that when an antiparticle encounters its particle complement, under certain circumstances, they mutually annihilate, releasing energy. Mesons are short-lived particles, and as such, the bonds that hold their constituents together are necessarily unstable. Presumably, then, one must also think of this binding not as a substance, but as a force. One generally imagines force as an invisible phenomenon that reveals itself through events in time—apprehension of it is indirect. One witnesses its effects but not force as a material substance. According to Lakoff and Johnson, in an object-event schema, "force is applied to the effect," whereas in a location-event schema, that same force is "applied to the affected entity" (199). In the former, since force is imagined as an immaterial, indirect effect, it would intuitively require a material source—imagined in the form of an agent, however abstract. I will return to this issue of agency in the model shortly.

To reiterate, the key kinesthetic image schema here is that of binding. In *The Cosmic Landscape*, Susskind writes of the creative process that produced his hadronic model.

> Actually, the word *string* is not what flashed into my mind. A *rubber band* is the way I thought of it: a rubber band cut open so that it became an elastic string with two ends. At each end I pictured a quark or, more precisely, a quark at one end and an antiquark at the other. (206, emphasis in original)

Such an association would seem to be appropriate: both rubber bands and strings are objects that in everyday experience do the work of binding. In fact, according to the *Oxford English Dictionary (OED)*, the etymological source of the meaning of the word *string* is the Latin *stringĕre*, "to bind or draw tight." But the image of a rubber band struck Susskind as being more apt for describing the dynamics of the internal mechanism of the meson. Stretched between the two images and perhaps shrinking from the image of the rubber band's overtly quotidian connotations, he seems to have settled on a hybrid image,[21] one that incorporates the advantageous attributes of both: "Le [*sic*] us suppose that a meson is composed of a quark-antiquark pair at the ends of an elastic string" (483). But I want to reemphasize here that the mathematics in no way demands a designation of string. There is no obvious indication that such "behavior" can only be that of a string. Susskind could have stuck with harmonic oscillator. Or, employing an alternative naming convention, he might just have well called this new theoretical object an Euler-vortex. After all, it was Leonhard Euler who developed the equation—the Euler beta function—that Veneziano used in 1968 to launch string theory as we know it today. And another theorist in Susskind's position may have decided that the image of a vortex was a more apt designate for the object, since a vortex is generally understood to be an extended object that rotates around an axis yet may also move through space.[22]

In English, of course, the word *string* can serve as both noun and verb. As a noun, its primary definition, according to the *OED*, is "a line for binding or attaching anything; normally one composed of twisted threads of spun vegetable fiber." Its secondary definition is "in an animal body: a ligament, tendon, nerve, etc.; an elongated muscle or muscular fiber." Both senses imply the work of binding or connecting. Strings are things that are both organic—existing in nature—and artificial, made by human hands. Strings can also be manufactured from inorganic substances, such as steel or plastic. Since the Renaissance, the word has also come to signify "a number of objects strung on a thread; hence, a series, succession." This shift in meaning represents the rudiments of its abstraction (in the conventional sense). The string comes to stand metonymically for the objects attached to it in an ordered succession. In the first half of the twentieth century, the term was further abstracted to function in techno-scientific discourses: in mathematics it has come to signify "a sequence of symbols or linguistic elements in a definite order"; in computer science, "a linear sequence of records or data."

The sense in which physicists use the term string clearly would seem to be more in keeping with this modern abstracted variation. For Susskind, the string stands metonymically for "a system of harmonic vibrations" that binds quarks together. But within the imaginary the image of the string exceeds its abstraction. While the image of the string is certainly a special, relatively vivid example of a harmonic oscillator, in the theory's original context—the dynamics of strongly interacting hadrons—the image of the rubber band felt, at least to Susskind, to be more appropriate because a rubber band is more elastic than the usual string. The tension in a rubber band increases as it is stretched apart. On the other hand, the image of the rubber band suffers from the disadvantage that it is perhaps *too* quotidian; it strains realism to project such an object into the atomic nucleus. Conversely, the image of the harmonic oscillator would seem to be too generically abstract, too verbose, too technical. The image of the string represents an ideal in terms of its general motility of signification. It can function dually as a concrete object—an object with a rich and long-standing domesticity—and as an abstraction, as a sequence of variables or symbols or bits, whatever the context may require. Moreover, it also has the advantage of being potentially both natural and artificial, both discoverable-in-nature and man-made—a critical feature with respect to the role that agency, both human and nonhuman, plays in the imaginary.

What these senses of the image of the string possess is a certain conceptual status that Lakoff would call "basic-level." In *Women, Fire, and Dangerous Things*, Lakoff defines basic-level categories as those that are "in the middle of a general-to-specific hierarchy" (13). He argues that basic-level categories are "functionally and epistemologically primary" and that they "are basic in four respects."

> Perception: Overall perceived shape; single mental image; fast identification.
> Function: General motor program.
> Communication: Shortest, most commonly used and contextually neutral words, first learned by children and first to enter the lexicon.
> Knowledge Organization: Most attributes of category members are stored at this level. (47)

Basic-level categories are basic precisely because we may readily imagine them. This ready imagining develops in large part because, with respect to objects in particular, they are a common and regular part of our daily experience. We frequently interact with them physically (what Lakoff calls "general motor program") and understand them in terms of their parts and the functions of the attributes associated with those parts. As a consequence, in our "knowledge organization," basic-level images hold a privileged, central position, unlike superordinate and subordinate categories (e.g., superordinate: animal, basic: dog, subordinate: golden retriever). As basic-level categories, phenomena such as doors, dogs, and digging strike us as unambiguously concrete.

Moreover, even ostensibly abstract phenomena such as the strings of string theory are readily imagined as real precisely because such technical terms depend on the conventionalized familiarity of basic-level categorization. In effect, when theorists venture a physical interpretation of mathematical argument by means of a term such as *string*, what they are invoking is not the "real world" per se, but rather an imaginary where a basic image drawn from conventionalized, embodied, human-scale reality "exists" within an enveloping space, also imagined as categorically basic.[23] A rich, varied, and long-standing array of connotations makes the image of the string instantly and viscerally real; its lexical status as basic conjures substance out of abstraction. In turn, the string's basic-ness buttresses its epistemological centrality within the abstract space of the imaginary.

To continue, within Susskind's imaginary, what were previously "operators" have now become a single "elastic string." The "space of possible states," which were also "rotational and vibrational states," transforms into a provisionally substantive object. A location-event schema has once again flipped into an object-event schema, where the string, having the attributes of extent along one dimension as well as elasticity, may behave in the expected ways; it may rotate and vibrate. Susskind then immediately adds an additional layer of intricacy to the imaginary.

> As the string moves in space-time a two-dimensional strip bounded by the trajectories of the quarks is generated. In analogy with Minkowski's world-line we call such a configuration a world-sheet. (483–84)

As an object situated in space, the string may "move" through that space, where time is imagined as an extra space-like dimension. This once again opens up an amphibology between object-event and location-event image schemas. In the object-event frame, a putatively substantive string "moves" through space; that movement reflects the passage of time. But with time imagined as a space-like dimension, that movement ceases to be insubstantial: it becomes something imagined as solid in its own right—a "strip." Susskind then dubs this "strip" a particular kind of "sheet." The notion of a strip or sheet exploits the convention in theoretical physics, originating principally with popular interpretations of Einstein's theory of general relativity, of imagining space-time as fabric-like.

Susskind explains that the movement of the string "generates" the "world-sheet." That act of generation not only produces a new abstract object, it also effects a shift from an object-event frame to a location-event frame, where the sheet has now become "world"-like. Space, time, and the objects contained therein become subsumed as attributes of an enveloping meta-space, the world-sheet. In this imaginary, one space-like image would appear to be "stacked" atop another, space-time itself: the world-sheet "sweeps" (as it is described elsewhere, playing off the schematic logic of the image) through space-time. Implicit in Susskind's argument is that this "world-sheet," in turn, stands metonymically for the entire "world" of mesons/hadrons. As an imaginary, this model world bears with it a constellation of traits, some more readily construed as concrete, others as abstract. Its complexity paradoxically both mimics the complexity of the familiar, human-scale world, also an imaginative *mélange*, while also contradicting it through its seemingly counterintuitive inferences and associations. In effect, Susskind conjures, more or less arbitrarily, a basic image—the string—out of a highly heterogeneous and rather incoherent imaginary.

Let us now consider the various agencies at play in Susskind's imaginary. At the outset of the introduction, Susskind states that "our goal in this paper is to explain how the dual-symmetric amplitudes *arise naturally* as the solution of an elementary quantum-mechanical model of mesons" (457–58, emphasis my own). The "dual-symmetric amplitudes" he mentions refer to the particular scattering patterns of mesons which, in this paradigm, are understood in terms of dual sets of vibrational states ("amplitudes") that emerge as a consequence of their internal structure. Within the exposition there lies an epistemological imaginary that structures and lends persuasive power to the argument by affording an imagined encounter between the theorist and the objects within the abstract theoretical space. That epistemological imaginary represents a confluence between a procedural space and an abstract theoretical space, between method and cosmos. It is an imagined meeting place, albeit highly abstracted, between theoretical praxis and the world in which that work has an impact. This is not simply the theorist presenting his own constructions as nature, of making the subjective, by sleight of hand, objective, but rather, it is a way of knowing the cosmos by imaginatively engaging with it in a way that complements indirect physical intervention through instruments made precise by mathematics.

The structure of this imaginary is essentially that of an abstracted journey with an origin, a path, and a destination, imagined in terms of an outcome or goal. The origin is the problem addressed—the "mysterious" and heretofore unsatisfactorily accounted-for behavior of mesons. The path is the ordered progression of the mathematical argument, which concerns itself principally with detailing the "mechanics" of a particular "model." Susskind presumes that if this model is correctly structured, that which it produces will cease to be an artificial mechanism. Rather it will allow for a certain phenomenon to "arise naturally." The imagining of construction will yield to, as the destination of the epistemological journey, an encounter with a natural phenomenon, which reveals itself as such through its heteronomous natural agency.

A critical epistemological assumption underlying Susskind's imaginary, which allows for the possibility of encountering that which "arises naturally," is to conceive of an event as an action. In *More Than Cool Reason*, Lakoff and Turner describe how the imagination tends to conflate events with actions. Firstly, an event itself is a form of abstraction—what are recognized as objects must be cognitively set off from the underdetermined background flux of the world and understood to be interacting with each other through time in accordance with recognized rules of causation. Many of our commonsense rules of causation are organized around what we come to understand as the basic components of human agency. Lakoff and Turner write: "every action is an event but every event is not an action—an event without agents can be understood in terms of . . . EVENTS ARE ACTIONS" (74). Normally, we attribute agency to humans: people, endowed with intentionality, take action in the world and effect change; they can cause events to occur. In *Women, Fire, and Dangerous Things*, Lakoff lists the "interactional properties" of "prototypical causation."

1. There is an agent that does something.
2. There is a patient that undergoes a change to a new state.
3. Properties 1 and 2 constitute a single event; they overlap in time and space; the agent comes in contact with the patient.
4. Part of what the agent does (either in motion or the exercise of will) precedes the change in the patient.
5. The agent is the energy source; the patient is the energy goal; there is a transfer of energy from agent to patient.
6. There is a single definite agent and single definite patient.
7. The agent is human.
8. a. The agent wills his action.
   b. The agent is in control of his action.
   c. The agent bears primary responsibility for both his action and the change.

9. The agent uses his hands, body, or some instrument.
10. The agent is looking at the patient, the change in the patient is perceptible, and the agent perceives the change (54–55).

Lakoff argues that:

> most representative examples of humanly relevant causation have all ten of these properties. . . . Billiard ball causation, of the kind most discussed in the natural sciences, has properties 1 through 6. Indirect causation is not prototypical, since it fails in number 3, and possibly other conditions. According to this account, indirect causes are less representative examples than direct causes. Involuntary causation is less representative than voluntary causation. (54–55)

While "involuntary causation" fails the complete litmus of conditions, it nonetheless bears with it the traces of human causation. The abstraction of causation to its least anthropomorphic expression still retains the basic imaginative structure of event as action—implying an agent, however theoretical. With events that do not directly involve people, one nevertheless may understand them in terms of an abstracted human action—where some agent effects a change in circumstance. This is a deep-seated, subtle, and highly conventional form of anthropomorphism in that events are understood in terms of the product of the action of abstracted agents, objects invested with a certain kind of agency that has been abstracted away from human agency, but nonetheless retains traces of its epistemological structure.

On first consideration, the investment of objects with agency seems trivial. To recognize and articulate events as actions is a commonplace, nearly universal way of structuring and thus understanding the world. Lakoff is not suggesting that events do not occur in the world, but rather, he contends, with Johnson, that "the topographic maps of the visual field, the orientation-sensitive cells, and other highly structured neural systems in our brains not only create image-schematic concepts for us but also create the experience of space as structured according to those image schemas" (*Philosophy* 509). While events undoubtedly do occur autonomously in nature, it is through their imaginations that theorists such as Susskind in their exposition project *in*to such events their recognizable structure as events, and the abstracted agencies by which they make sense of them.

To state an obvious yet nonetheless important point: theorists simply do not recognize events that they cannot imagine as possible. And that which is possible is necessarily structured by the interactional properties made manifest through cognition. It ought not be taken as a given that we are able transparently to recognize any and all non-human agency in the world, a cosmos independent of our own projective anthropomorphisms, however obscured. When Susskind writes that a "string moves" and that this movement causes a "two-dimensional strip" to be "generated," among the myriad imaginative assumptions made, he is also taking as given items 1–6 in Lakoff's list of "interactional properties" for "prototypical causation." Furthermore, in the imaginary, he is then able to encounter and thus bear witness to this event and to subsequently "call such a configuration a world-sheet." This blending or encounter of human and natural agencies within an imaginary such as Susskind's is, in this respect, something wondrous—that we are able, by the imagination, to exceed the imagination and summon forth the natural. Susskind's imaginary manifests the intuitive and the counterintuitive as productively coextant.

My use of the notion of agency here is compatible with what Andrew Pickering calls "material agency." For Pickering, the "interdefinition of human and material agency" constitutes what he calls the "mangle" that is scientific practice (*Mangle* 25). That scientific practice is a mangle speaks to its radical heterogeneity and, in terms of an imaginary, the complex ways in which imagery, mathematics, and embodied praxis interact. My complication of Pickering's formulation is in insisting on the pivotal role that the imaginary plays in this constitutive intertwining. A definition of human and material agencies must necessarily possess both a practical and an imagined component—prompts for action, a form, as well as substantive content. Susskind's imaginary reflects a tension between these intertwined human and material agencies. The construction of objects, of mechanism, of system interpenetrates a "physical interpretation," an imaginative invitation for material agency to "arise naturally."

One of the crucial ways that the kind of technical exposition under scrutiny here purifies itself of subjectivity—thereby imaginatively facilitating an encounter with objective nature—is by systematically, almost ritually, absenting overt human agency from the imaginary. For example, Susskind asks us to "suppose that mesons *are composed* of bound" quark-antiquark pairs (459). This is a suppositional or speculative imagined space where no one in particular composes the model, but rather, by means of the passive verb, mesons uncontroversially "are composed." Initially in the model, an abstract, absented human agency, in constructing the model within a space of supposition, binds quark pairs together. But ultimately, such a model "is bound," not by an imaginative act, but "by purely harmonic forces" (460). While following the conventions of technical nomenclature, Susskind sets "operators" going within the model, which in turn and of their own accord, "act" such that they "generate" those very harmonic forces (in the form of "rotational and vibrational states"), in keeping with this imaginary.

In the second passage examined, Susskind repeats: we "suppose that a meson is composed of a quark-antiquark pair" (483). But in this variation on the imaginary, forces-as-vibrational-states are substantiated as "an elastic string." No one, though, sets the string in motion; it "moves" of its own accord, and as a result a "world-sheet . . . is generated" (484). Admittedly, the encounter with natural phenomena implicit here is highly obscure; it comprises an anthropomorphism abstracted away to a certain economical minimum. In keeping with Lakoff's definition of "prototypical causation," a natural agent is an agent that acts in such a way as to partially satisfy the complex of characteristics recognized as human agency for an event-as-action schema. In this case, the string plays the role of natural agent; its motion generates a world-sheet. But, to reiterate, the conformity to prototypical agency is partial: among other issues, there is an inherent ambiguity between agent and patient, since the string is, in the one imaginative frame, part of the world-sheet, and as such, not definitely discrete.

Nevertheless, one is meant to acknowledge the *world*-sheet as just that—as the world, in all its natural objectivity. In the logic of the epistemological imaginary, the string emerges as a natural, heteronomous agent within this particular world. The string exists because the world it generates exhibits a satisfying richness of necessary detail. By means of this imaginary, Susskind posits an analogy where image is to concept as mathematics is to reality, since mathematics and concept are equivalent. Image doubles for concept just as concept doubles for reality. Yet, surreptitiously, a third relation informs this analogic textual structure; namely, that of the images—quark, string, and world-sheet—to an imaginary, the microscopic world recognized in rescaled, recombinant human-scale terms. It is the imaginary that then doubles for reality—and as consequence, grants it its argued-for substantiality. It is perhaps at this moment in string theory's infancy that the string "arises naturally." However, with respect to its epistemological status, the string remains provisional, embedded as it is, within a heterogeneous imaginary that contradicts a commonsense physical coherence.

The Heterotic Superstring

A second example of a string theory imaginary comes from "Heterotic String Theory (I)," written by the four theorists—Gross, Harvey, Martinec, and Rohm—and published in 1985. This article represents what many string theorists view as critical groundwork laid for the subsequent development of the most "realistic" version of the five superstring theories that emerged from the productive ferment of the "first superstring revolution" of the mid-1980s. Theorists are inclined to designate a new model "realistic" when that model reproduces the physics of the standard model of quantum physics. Most theorists thus consider the heterotic string, while the most realistic of the superstring theories, only "semi-realistic," in that it produces a close approximation of the standard model.[24] But as we shall see, that a given string theory is "realistic" (in contrast to the now discredited hadronic string) in no way ensures that it is also imaginatively coherent.

At the beginning of the paper, Gross et al. provide a summary description of the "heterotic" superstring.

> The heterotic string is constructed as a combination of the right-moving coordinates of the ten-dimensional superstring and the left-moving coordinates of the 26-dimensional bosonic string. This yields a theory which is Lorentz invariant in $D = 10$ and which has appealing features not found in either theory separately. . . . Thus the name "heterotic." (254–55)

In short, a superstring theory such as that of Gross et al. maps points on a world-sheet, which correspond to the center mass of a string, into an abstract multidimensional space that represents space-time. It is "Lorentz invariant" in that it is consistent with special relativity, a precondition for any realistic physical theory. Also, unlike the hadronic string, this particular version of the string is "super-" because the model incorporates supersymmetry, a theoretical structure that, in pairing both fermions and bosons with their hypothetical superpartners, allows for the synthesis of the standard model's three distinct force gauge groups into one "supergroup." One consequence of this "super" reformulation of the string is that it now has as its fundamental scale the Planck length. Another consequence is that for mathematical consistency, the superstring must reside in a space-time of ten dimensions: the D in the passage stands for dimension. The position of the heterotic string is "embedded in ten-dimensional flat space-time" (256).

Gross et al. chose the qualifier *heterotic* (from *heterosis*), exploiting the word's sense of "different" or "mixture," since their new model joins 10-dimensional fermionic superstring theory with 26-dimensional bosonic string theory.[25] The idea here is that while the entire theory is embedded in ten dimensions of space-time, there is also another level, so to speak, of structure. This is what I will call the theory's *attribute space*—an abstract space that bears with it the attributes of the strings being treated. It is this attribute space, with its own dimensionality, that is "embedded" within space-time. In the heterotic superstring imaginary, we find one abstract space, attribute space, coupled to and interacting with another abstract space, space-time.

Later, Gross et al. call their new theory an "amalgamation of the old fermionic and bosonic strings" (282). They also state that the "new string theory is constructed as a chiral hybrid of these" (254). Chiral here refers to the handedness of certain particles in quantum theory, in particular, certain weak force bosons, which are effectively left-handed; their weak charge spin has a left-moving direction. Recall from Veneziano and Susskind that string vibratory patterns are "dual" in the sense that they have distinct left-moving and right-moving components. Chiral particles exhibit characteristics that are asymmetric in that they are not superimposable on their mirror images in attribute space. Of course, they are not left-handed in the everyday literal sense. In fact, to designate coordinates as left-moving or right-moving is motivated by, in certain respects, imaginative convenience and convention. It takes advantage of the ready familiarity we have of notions of left- and right-directedness as abstractions predicated on human handedness. Such a distinction—movement in opposite directions—could just as easily be termed back- and forth- moving, or up- and down- moving. Left and right have meaning in relation to human bodies situated in a space conceptualized as having three dimensions—in contrast to the terms we use to designate the other two dimensions of possible embodied movement (e.g., up and down, and back and forth). This becomes starkly clear when we attempt to imagine ten or twenty-six such dimensions of space in which a hypothetical body could potentially move. In a ten-dimensional world, one would have to invent six additional sets of orienting terms in order to give directions for navigating the space successfully.

Shortly thereafter, Gross et al. further explicate the features of their new theory.

> We shall construct in some detail the free heterotic string. In sect. 2 we present an operator construction using light-cone gauge quantization. Much of this is simply putting together the bosonic and fermionic components. The new feature is the treatment of the 16 internal, left-moving, bosonic coordinates. We show that these must, for consistency, be compactified on a particular torus. This compactification leads to a set of 496 massless vector bosons. (255)

Dividing the twenty-six left-moving bosonic coordinates into two groups, they designate sixteen of these as "internal," implying that the remaining ten are thus external. Here is the image of a space within a space within a space. An "internal" container-like space of sixteen left-moving bosonic coordinates is embedded within a container-like attribute space that couples yet leaves distinct the remaining ten left-moving bosonic coordinates and the ten right-moving fermionic coordinates. This imagined space, in turn, is embedded within a ten-dimensional abstract space of "flat spacetime." Since the external space of left-moving bosonic coordinates is ten-dimensional, each coordinate within that space contains a supplementary container-like "compact" sixteen-dimensional space. Gross et al. describe this supplementary space as having "hidden, compact spatial dimensions," and are concerned with "the structure of the internal sixteen-dimensional space" (253, 258). That structure must, "for consistency, be compactified on a particular torus." A torus is a donut-like geometric object defined mathematically by the notion that some spatial dimensions are flat while others are positively curved.[26] So the imaginary, in addition to placing a "hidden" space within a space within a space, is further complicated by a multiple multi-dimensionality that is not only imagined as flat space but also as curved space, a "compactified" torus of sixteen dimensions. Again, we notice a particular blending of object-event with location-event schemas. A space, normally a location recognized as ground in a figure-ground dual, becomes an object, and thus the figure, embedded within yet another space, a ground. Figure and ground repeatedly shift places, and thus by necessity, transform in scale and interactive capacity within the imaginary. Mathematical consistency yields a radically heterogeneous, if not completely baffling, imaginary.

Later in the article, Gross et al. state that "this compactification produces 16 gauge bosons associated with the isometries of the torus, but in addition yields 480 massless solitons which wind around the torus" (282).[27] The first 16 "gauge bosons" are aspects of the space itself, of the "isometries of the torus," but the remaining 480 particles that the act of compactification "yields" actively "wind around the torus." Rather than aspects of the space itself (a location-event schema), these 480 solitons possess attributes—they wind, more in keeping with an object-event schema. The heterotic string's status as "semi-realistic" stems from this result—that the string generates a particle spectrum that includes necessary gauge bosons, but also additional particles not stipulated in the standard model. In terms of the imaginary, this passage is significant in that it piles yet another abstract space into the mix—where some particles are specific locations in that space, while others, as objects, "wind around" the compactified sixteen-dimensional torus.

Gross et al. state that their new heterotic superstring is exclusively a closed string, not an open one (a loop as opposed to open-ended). These closed strings "are constructed" by "quantization of an action" given by the "area of the world sheet swept out by the string" (255). By making the closed string consistent with the "light-cone gauge . . . the theory reduces to a two-dimensional free field theory" (255). So, in addition to the abstract spaces of ten-dimensional left-moving bosons, ten-dimensional right-moving fermions, sixteen-dimensional left-moving bosons compactified on a torus, and ten-dimensional space-time, one is called upon to imagine the first three of these theoretical objects "reducing" to a two-dimensional field, a world-sheet that "sweeps through" space-time.

My point here is not to call into question the science, but rather to highlight the extraordinary nature of the imaginary that accompanies what are consistent mathematical arguments. The imaginative achievement is in the retrofitting of basic human-scale images—string, dimension, space, and implicitly, basic spatio-temporal kinesthetic relations such as inside and outside, left-moving and right-moving, action-as-event, linking, container—to a domain where there can be no expectation that such phenomena as they are generally known and recognized on human scales would at all be pertinent or viable. The impression of the utter weirdness of such imaginaries is very much a product of their heterogeneity—the medley of abstract space within space, object within object, where no schema in particular, whether principally based on an object-event structure, location-event structure, or otherwise, would seem to stand out as the one monologically coherent cosmic order. As an imaginary, the cosmos described by string theory technical exposition becomes a complex of multiple and often disparate images.

As we saw with Susskind, Gross et al. use the passive of the verb *construct* in their introduction: "the heterotic string *is constructed*" (254). This once again reflects a pervasive rhetorical strategy in string theory technical exposition that seems to be a relatively unambiguous scientific convention, but tellingly, in terms of an imaginary, makes use of a particular form of imaginative absenteeism of human agency. The narrative pattern proceeds thus: theorists project their proxies into abstract space(s) in order to construct something which, if assembled properly, works—in the sense that it allows for a summoned phenomenon to "arise naturally," to repeat Susskind's expression. With abstracted human agents present yet absent, a string "is constructed" such that "this yields" a theory which has "appealing features." An artificial process provides for the possibility of encounter with natural phenomena that are "appealing" precisely because they exhibit features that are physically realistic; that is, natural. Construction is not presented simply as nature, but rather, it "yields" a space of genuine discovery beyond its own machinations. In this instance, the qualifier "heterotic" suggests that the appealing features are loosely analogous to the way that crossbreeding yields a more appealing or advantageous crop or domestic animal. The theorists project their imagined proxies into an abstract space, manipulate its contents, and then tacitly invoke yet another abstract space—an epistemological space that enables their manipulations to yield that which is not constructed but rather autopoietic, self-generated.

This rhetorical device recurs in a slight variation immediately thereafter: "we shall construct" (255). Still in keeping with stylistic convention, Gross et al. use the active, formal future of the verb, a tentatively more emphatic declaration of agency within the abstract space. But in the next sentence they subtly distance themselves again from such agency: they "present an operator construction." While "operator" denotes for the theorists a specific mathematical procedure, in the imaginary, it connotes a sense of agency in its own right, an operator literally being one who does or effects something. This work, they contend, "is simply putting together . . . components" (255). Such a statement again partially absents their own agency from the abstract space. The imaginative pattern becomes clear: construct, withdraw, then yield to the discovery of a heteronomous agency. In this case, their compactifying the sixteen "hidden" dimensions "leads to" the discovery of a "set of" bosons—quantum particles, which are natural phenomena. That which was previously hidden, they "show" to be true. Just as there are a profusion of hybridities at stake here—the commingling of images and their concomitant schemas to serve the requirements of the expository argument, there is also an implicit encounter between imagined human and non-human agencies. By framing their capacity to do work—a procedural space that supplements mathematical technique—in terms of an encounter within this abstract theoretical space, the theorists further complicate the imaginary at stake. Once again, what was alien, remote, and utterly counterintuitive, beyond the grasp of human-scale imaginative resources, becomes quasi-domesticated. The space they delineate is made coherent through a conventionalized graspability, an implicit willingness to look past the exposition's imaginative inconsistencies.

The encounter of human and non-human agencies, in turn, corresponds to an epistemological imaginary, one that exploits the imaginative conceit of reality as that which lies beneath a surface—and that which must be revealed. I call "reality-as-depth" an imaginative conceit because it is, to quote Lakoff and Johnson, an "image plus knowledge about the image plus one or more [inferential] mappings" (69). An imaginative conceit such as "reality-as-depth" incorporates abstraction while nonetheless foregrounding a basic image that provides the commonsense comprehensibility for that abstraction. In this case, the conceit of reality-as-depth reinforces the substantiation of string-as-depth, as fundamental substance that lies within or below the form of the real world, and out of which quantum particles—whether bosons or fermions—manifest. In their knowledge, the theorists may mark the heterotic string as the deeper truth. The fundamental conceits of form and substance transform in a way akin to the reversals of figure and ground discussed earlier. In one imaginative frame, string is the substance out of which are formed quantum particles. In another, the string itself is the form, and its substance remains a mystery, the deeper truth still hidden. In their concluding remarks, Gross et al. concede:

> It is certainly true that the formulation of the heterotic string appears somewhat awkward and contrived. This is not a shortcoming of our theory; rather it is indicative of the present level of understanding of all quantum string theories, which leaves much to be desired. Many of the most remarkable features of these theories emerge without a full comprehension of their origin. Most mysterious are the general coordinate invariance and local gauge symmetries of string theories, whose appearance lacks a geometrical explanation. This suggests that there exists a more profound formulation of string theory, in which these features would be manifest. In such a formulation the heterotic theory might appear more natural. (282)

I find this defense of the theorists' version of string theory remarkable for a number of reasons. Implicit in the admission that their formulation "appears somewhat awkward and contrived" is a conflation of imaginary with conceptualization (through mathematics). If a given mathematical argument is self-consistent and correct, how can it be awkward and contrived? This admission further makes use of, once again, the epistemological binary of appearance and reality, where appearance is "awkward and contrived," whereas, by implication, reality would be the opposite—elegant and coherent. But is there any *a priori* reason why reality—as it is imagined—decidedly should *not* be awkward or contrived? The theorists blame this "shortcoming" of the theory not on their own formulation, but on the "present level of understanding of all quantum field theories." Again, a future "level of understanding" would presumably go "deeper." Immanent in their defense is the imaginary that physical reality—and our capacity to understand it—consists of "levels" and that a "more profound formulation" would somehow capture a deeper reality, and accordingly, "might appear more natural." But any "deeper" understanding in the form of better mathematical arguments, ones they mark as elegant and coherent by a standard that does not translate to exposition, may very well yield only more incoherent and heterogeneous imaginaries.

Gross et al. continue to take advantage of the structure manifest in an image schema of reality-as-depth as they further elaborate on an image of the cosmos. They state that "many of the most remarkable features of these theories emerge without a full comprehension of their origin." By "origin," the writers presumably mean a more fundamental formulation of "these theories" where, as in general relativity, first principles lead "naturally" to descriptions of the physical phenomena at stake. The use of the word "origin" here reinforces the sense of that which arises from a source, where nature consists of layers whose clearing away reveals further degrees of "comprehension" of the truth. The overlapping of images of emergence from an origin and reality-as-depth plays off an imaginary traditionally linked to the basic-level and conventionalized phenomenon of a spring. Here then we have another case of a "natural" image abstracted (quite far) away from its human-scale origins, but nevertheless retaining its affective persuasiveness. There is appearance, and beneath it, reality. Appearances are "mysterious" in that they obscure their origins, which the theorists assume originate in "a geometrical explanation." This "more profound formulation . . . might appear more natural" in that construct would yield more readily to discovery. Since geometry plays a fundamental role in general relativity, Gross et al. seek first and foremost "a geometrical explanation" in order to make their theory consistent with general relativity, given its status. Nevertheless, in terms of affect, the "geometrical explanation" would be the candidate most likely not to strike them as awkward and contrived—to be the most emotionally satisfying, since geometry, of all the mathematical formalisms, presumably is the most intuitive, the one most readily imagined as elegant and coherent. Yet the satisfying intuitiveness of geometry, its potential for elegance and coherence, generally speaking, may well be a product of the close association in the pedagogical culture of theoretical physics between it and a host of human-scale, familiar examples—examples that lend geometry, in particular, a substantiality not enjoyed by other formalisms. The potential for geometry to be the most elegant and coherent "formulation" of physical reality may also be a matter of convention; it has that potential simply because the community of theorists marks it as such.

What one must contend with is that as mathematical arguments within string theory technical discourse become ever more complex, an imaginary that accompanies theory may become all the more counterintuitive and bizarre. I offer the following analogy to help imagine the process by which theorists manage the three fundamental components of their theorizing, which I have referred to throughout as mathematical procedures, imaginaries, and embodied praxis (experiments), in order to produce string theory. All are necessary to string theory as scientific practice, and together they are mutually constituting. Mathematics provides precise prompts for action—constraints that guide but do not represent.[28] As an essential complement, an imaginary affords a means of sharing among practitioners an experience of remote spaces and alien encounters—the substantive content to mathematical form. An imaginary mediates between body and world and grounds mathematics in a community with its own particular social conventions; that is, to a culture. This culture structures both affect and relationship—what to feel about the abstract spaces, how to feel it, and how to interact with the phenomena that populate those spaces. These three components of, in this case, heterotic superstring theory—mathematical argument, an imaginary, and experiment—are all indispensable.

Together they effect a kind of triangulation, and in so doing, function in a way analogous to a high-powered microscope.[29] A theorist "sees" a virtual medium constructed through a series of feedback loops from the intervention of abstracted hands with putative objects in a remote space. In the analogy, the probing particle beam of the microscope is akin to prosthesis: it allows access by a human body to remote spaces on extreme scales. Mathematics, in turn, allows for the precision necessary to make the instrument sufficiently sensitive to interact effectively with these remote objects, ultimately allowing for the proper discernment of their behavior. And the computer screen, which presents digital data as analogic illustration, is akin to the imaginary—that through which the theorist "sees" the so-called objects being probed.

This relationship between mathematics, an imaginary, and praxis helps to explain why the further theorists "delve into nature," the more counterintuitive and just plain weird it seems. There is absolutely no *a priori* reason objects and events that have gained their epistemological status as stable phenomena from commonsense experience, from human bodies and minds interacting with the world on human scales, where this experience is further conditioned by historical and cultural contingencies, ought to obtain on scales far removed from those scales. Lakoff and Turner pose a question that is entirely relevant to string theory technical discourse as scientific knowledge: "How can we manipulate our conceptual resources in such a way that we can create ways of understanding other things in terms of ourselves?" (74). Such a redeployment of "conceptual resources" allows string theorists to understand the "things" about which they are concerned not simply in terms of "ourselves," but more specifically, in terms of the full cache of human-scale objects, events, and spaces, however abstracted from their origins, available to their imaginations. It is a testament to the facility of the imagination that string theorists are able to recombine, rescale, and redeploy images that originate at or near this "home" to far-flung spaces and make them at all capable of facilitating repeatable, effective physical intervention in those realms.

It would seem that the heterotic superstring may teach us this: one can only expect that the further theorists push back the boundary of the known into the unknown, the imaginaries through which they mediate that encounter with natural origins will become ever more hodgepodge—ever more, as Serres puts, "a mixture, tiger-striped, motley, mottled, zebra-streaked, variegated" (*Genesis* 111). This radical heterogeneity, however, is not uniformly stochastic: patterns, logics, and regularities may organize local structures. There is, nonetheless, no ironclad law that declares that a given imaginary that affords one entry to remote spaces should in any way be coherent or unitary (even if an unmediated world might be). In the case of string theorists, they blend images in ways that they, by consensus, agree are appropriate, if not imaginatively coherent. Part of the imaginative achievement of this work is in its refusal to adhere to human-scale expectations of coherence. And ironically, it is perhaps the theorists' longing for a coherent "theory of everything"—a "master equation," as Brian Greene puts it, that provides the impetus to cobble together an ever more disparate imaginative heterosis—with all its hybrid vigor (*Elegant* 5). Newtonian cosmology, as it traditionally has been understood, strikes us as unambiguous because we mark it as such—and furthermore, because it simply excluded so much.[30] To the extent that an imaginary that prescribes the heterotic superstring is more heterogeneous than that of the hadronic string, such relative incoherence speaks to the prospect that the more a theory endeavors to incorporate ever more detail concerning the "real world," the more of a farrago it would necessarily become. As mentioned in the introduction, theorists often complain that the standard model feels like an unsatisfactory kluge, cobbling together as it does the three subatomic forces, while requiring numerous constants (such as rest mass) to be plugged in, and excluding, as it does, gravity. While the heterotic string, on first glance, simplifies the disparate quantum particles of the standard model into one core image—the string, that image itself has no relevance outside of a complicated bricolage of yet more images. A push to elegance and cohesion displaces elsewhere the counterintuitive complexity requisite for ostensibly representational exposition to match the intricacies of mathematical argument.

The Braneworld Model

As mentioned earlier, Randall and Sundrum's "An Alternative to Compactification" was an important and influential paper within the field when it was published in 1999. It exemplifies the kinds of modeling being done in the period following the "second superstring revolution" and is one of the first and most convincing attempts to propose a cosmological model where the extra dimensions required by string theory are not microscopically compact, but rather, extended out into cosmic scales. This is a cosmic model where the known universe of four dimensions exists in a multidimensional bulk or metaverse. While Randall and Sundrum do not consider themselves string theorists per se, their work on braneworld models relates to string theory in that it follows on the mid-1990s work of the likes of Ed Witten and Joe Polchinski and takes up a problem central to string theory: the reconciliation of gravity with the standard model, using a theoretical object—the brane, that is also pivotal to M-theory. In their paper, Randall and Sundrum offer an alternative to the "lore," as they put it, "that convinces us that we live in four non-compact dimensions" (1).[31] Earlier versions of superstring theory attempt to reconcile their extra dimensions with gravity by compactifying them, since "Standard Model matter cannot propagate a large distance in extra dimensions without conflict with observations" (1). Randall and Sundrum remind the reader that while this conflict with observations "can be avoided if the Standard Model is confined to a (3 + 1)-dimensional subspace, or "3-brane," in higher dimensions . . . this solution will not work for gravity, which necessarily propagates in all dimensions as it is the dynamics of spacetime itself" (1). They argue that this "story can change significantly" when the assumption "is dropped" that the extra dimensions must be compact (1). In effect, "we can live in 4 + *n* *non-compact* dimensions, in perfect compatibility with gravity" (1, emphasis in original).

The paper concerns itself with enumerating this new model. Randall and Sundrum write:

> The set-up for our theory is a single 3-brane with positive tension, embedded in a five-dimensional bulk spacetime. . . . we choose to first work in a finite volume by introducing another brane at a distance . . . from the brane of interest, and taking the branes to be the boundaries of a finite fifth dimension. We will eventually take this second brane to infinity, thereby removing it from the physical set-up. (2)

The pivotal image here is "a single 3-brane." On the one hand, one is to imagine this object as a container within which is to be found a particle spectrum that, presumably in future formulations, would accommodate the standard model. This 3-brane that contains a certain particle spectrum serves metonymically for the known universe. It is in this sense that a 3-brane can be a "braneworld."[32] This 3-brane is a "subspace" in that it is an abstract space within an enveloping or superordinate metaspace. The notion of a subspace derives from the image of the bulk as a container. A subspace then is to be imagined as a container within a container. On the other hand, one is to imagine the 3-brane as a membrane, in that it possesses the attribute of "positive tension," like a drumskin, on and as the boundary of a higher-dimensional space (i.e., akin to a two-dimensional drumskin on the boundary of a three-dimensional drum). Since it has tension, the 3-brane also vibrates in a way loosely analogous to the vibrations of a drumskin. As with the string of string theory, the vibrations of the 3-brane generate a spectrum of "Kaluza-Klein" (KK) particles, as well as the graviton.[33] The image of container and membrane are compatible in that a three-dimensional membrane (for example, the wall of a biologic cell) may play the role of a part of a container—as its boundary, demarking interior from exterior.

Graviton particles may pass between the brane and the bulk, but matter may not. According to the *OED*, the word *membrane* originates in the sense of "that which covers the members of the body." A membrane provides an exterior surface for a living body, and as such, is organic. In its history, membrane also has come to designate inorganic and/or inanimate permeable containers; for example, the drumskin. The image of the brane, imaginatively stripped from its organ, becomes abstracted from its organic, corporeal source. Etymologically, it occupies a mediate symbolic space between the mechanical and the organic, that which is made and that which makes itself. But the 3-brane in Randall and Sundrum's model complicates this image of membrane as covering. A brane is a contiguous object; its very substance constitutes the boundary. Matter particles are free to "propagate" throughout the brane; they are not stopped by a membranous boundary per se, but are simply bound to the substance of the brane. In this counterintuitive image, the brane is a membranous surface only in the specific sense that it has fewer dimensions than the object, the bulk, which envelops it. As in Gross et al. and Susskind's models, the imaginary vacillates between object-event and location-event schemas. In the object-event imaginative frame, the brane is a container with a boundary that "traps" particles within its interior while graviton particles are free to escape to the surrounding bulk. In the location-event frame, KK particles are the vibratory "spectrum" of the 3-brane as a dynamic abstract space. Conversely, graviton particles are topologic-like attributes of the abstract space of the bulk. Depending on the context, one can imagine the graviton as either a loop-like particle passing through the brane into the bulk, or as the vibration-like behavior of space-time itself, imagined as pool-like or fabric-like.

So, to reiterate, the 3-brane, both container and semi-permeable membrane-like space, is "embedded" in yet another container, a "five-dimensional bulk spacetime." In the bulk, one of the dimensions corresponds to time, leaving four dimensions of space, three of which also constitute the 3-brane. Therefore, in Randall and Sundrum's model, there is one supplemental spatial dimension—a "finite fifth dimension," which defies ready imagining. They introduce "another brane" at a specified distance from the original 3-brane, and together these two 3-branes form opposing "boundaries" of the bulk. So while the 3-brane itself is a container, it also functions as the boundary of yet another container, the bulk, the universe extended into another dimension to form a metaspace. Inasmuch as the 3-brane represents the universe, one can construe this bulk metaspace as a "metaverse"—that which contains multiple universes. The aptness of designating a theorized metaverse as a "bulk" hinges on the word's sense of heft and girth, of a large container holding many things. Like string and brane, the bulk, as a word that comprises a linguistic category, is cognitively basic—monosyllabic, readily graspable, and "in the middle of a general-to-specific hierarchy" (Lakoff 13). But the word has a complicated etymology. Archaically, bulk could refer to a body-trunk, thorax, or belly—of a body-cavity that holds organs. In other contexts, it could mean cargo or heap, of a cavity built to store goods. As with the image of the string, this dual sense lends itself to the conflation in a braneworld imaginary of human construct and heteronomous, organic-like natural whole.

Randall and Sundrum's reformulation of the problem of gravity in a braneworld model is to "trap" gravity not within the 3-brane but near it (1). The use of the word trap implies an object-event frame where gravitons are tiny particles limited in their movement to a bounded space. They argue that "a curved background can support a 'bound state' of the higher-dimensional graviton, which is localized on the extra dimensions" (1). Here the image of trapped gravitons is complicated by a sense that they are tethered (abstractly) to the space as a kind of ground, a "curved background," which supports the gravitons in a "bound state." The boundary that traps the gravitons is at once a feature of the gravitons, a "bound state," and also a feature of the space itself, something that the space "supports." The result is that "the graviton is confined to a small region within this space" (1). The "bound state" to which they refer "falls off rapidly away from the brane" (2). In keeping with the figure-ground reversals at play above, the imaginary here subtly blends schematic logics that are both complementary and at odds. In the first instance, the "curved background" serves as an abstract space, a ground that "supports"—in the sense that it carries or bears up, abstractly—the figure of the graviton. The most pertinent attribute of this object, the graviton, is that it is in a "bound state." As a consequence, the graviton "is confined to a small region within this space": as an object, it is bound to the abstract ground. But since the graviton is also the curved background itself, this bound state is not only an attribute of the object-as-figure, it is also a local ("localized") feature of the location-as-ground. As such, the bound state itself "falls off rapidly from the brane," suggesting an image of a precipice.

The most noteworthy feature of their new model, Randall and Sundrum contend, is in the fact that this "bound state mode reproduces conventional four-dimensional gravity" with "only a small correction" (2).[34] In this imaginary, then, unlike four-dimensional general relativistic space-time, the "curved background" has an extra, "fifth" dimension. But Randall and Sundrum are able to treat these "higher-dimensional gravitational fluctuations" in such a way that they reduce to "four-dimensional . . . states" (1). This procedure results in a graviton (imagined, conversely, as gravitational fluctuations) and a "tower" of supplemental "modes" (2). It is this tower of supplemental modes that gives the "small correction." On the one hand, one is to imagine gravitons in bound states, along with an abstract "tower" of particle modes (what they shortly thereafter refer to as a spectrum) sitting atop, in a sense, a dynamically curving ground. Yet this tower of modes and "gravitational fluctuations" are also aspects of the ground itself as the ground state, the state of lowest energy. This is counterintuitive precisely because we expect the imaginary that corresponds to the mathematical argument to be both coherent and, in some sense, unitary. Instead, what we get is a constellation of images with paradoxical schematic logics.

Like Susskind and Gross et al., the imaginary manifest in "An Alternative to Compactification" not only allows for the mixing and radical rescaling of often contradictory human-scale images, it further provides an abstract, epistemological space of encounter between imagined human and natural agents. There exists a delicate interplay between an inventive and intervening proceduralism and discovery framed in terms of an act of witnessing. Randall and Sundrum's proxies in this abstract space begin with a *set-up*—a standard term that, within the imaginary, concomitantly serves as an abstract object marked as something manufactured (2). They "choose to first work" with one particular configuration of this set-up (2). This work consists of "introducing another brane," then "taking the branes to be . . . boundaries" (2). They then "carefully quantize the system" (2). And after several other procedures, they "take the second brane to infinity, thereby removing it from the physical set-up" (2). The agency in this last image is striking: it is an abstracted leading or shifting ("take to") of the second brane to "infinity," imagined as the destination of a journey, however short. An implicit boundary marks the distinction between infinity and "the physical set-up."

Infinity here is imagined, in contrast to Susskind's hadronic string theory, not as a container exactly, but rather as a location in the epistemological metaspace of encounter; it is the exterior region delineated by the physical set-up as a container-like location. Furthermore, theoretical procedures within this epistemological space are imagined as a path with an origin, trajectory, and destination. These abstracted manipulations and ambulations ultimately result in "the existence of a bound state" (1). On the one hand, "existence" here has a precise meaning within standard mathematical language: for an equation with certain conditions, certain solutions will exist. Conversely, though, the choice to describe these "bound states" as something that "exists" here evokes an imaginative conceit of encounter with natural phenomena, since in this context, such an existence runs counter to a purely human agency. Since the bound state exists, it is not constructed, but rather "can be understood" (1). It "mimics pure four-dimensional gravity," and thus can co-opt its empirical authority, its natural-ness. Accordingly, this allows Randall and Sundrum to confidently declare that "we *can* live" in this particular braneworld model, a model paradoxically made real through its explication as an imaginary (1, emphasis my own).

That we can "live" in their braneworld model grants it, as an imaginary, a visceral reality. The impression that their set-up, once "appropriately tuned" (an image that plays subtly on the imaginative regime of musical notes, radio frequencies, and the tuning of instruments and radios), generates "a sensible effective four-dimensional theory" is further reinforced in the imaginary in that Randall and Sundrum label the two branes at the outset of the procedure "visible" and "hidden" (7, 3). They then explain that the "graviton is 'bound' to the visible brane" (3). Once they "take the second brane to infinity" (the "hidden" brane), they effectively remove "it from the physical set-up." Yet another epistemological assumption then is that what is hidden does not exist; it is unphysical. This allows the "set-up," in turn, to become physical; that is, real. Its reality is confirmed tacitly by the fact that it is "visible": we are able to bear witness to its existence by means of tests and the anthropic confirmation that we can (and perhaps do) live in it. As such, once again, Randall and Sundrum assert that "we can consistently exist with an infinite fifth dimension, without violating known tests of gravity" (7). The imaginary effects an Escher-esque imbrication of intervention and witnessing such that it gains a substantive reality in accord with human-scale experience—once again, an imaginative achievement. Model builders such as Randall and Sundrum have essentially extended the braneworld imaginary in such a way that human-scale agents are able abstractly to both perceive and manipulate entire universes as container-like objects suspended, so to speak, within a multidimensional metaverse. The boundaries of what is effectively imaginable within string theory have expanded to not only encompass within a visualizable horizon, but shrink to manipulable scales, entire universes.

The Romance of Encounter

Although the imaginaries presented by the three technical articles this chapter has examined do draw on images from a human-scale world, in order to accommodate the particular characteristics of these remote domains, the cobbling together of imagery winds up evoking a representation of the physical world with unexpected or counterintuitive properties. Such imaginaries are counterintuitive because, although they incorporate abstracted human agents and human-scale objects behaving in ways loosely consistent with human-scale, earthbound interactions, to double efficaciously for mathematical arguments, the imaginary must allow for radical juxtapositions, transformations, and interchanges that often defy commonsense or intuition. This is an imaginative taxonomy comparable, in some respects, to Borges's "Celestial Emporium of Benevolent Knowledge."

The strangeness of an imagined space in string theory, then, is in its lack of indifference, in the conflation of the contingently human and the objective. As Bachelard contends, "Space that has been seized upon by the imagination cannot remain indifferent space subject to the measures and estimates of the surveyor. It has to be lived in, not in its positivity, but with all the partiality of the imagination" (*Poetics* xxxvi). By "positivity," Bachelard is referring to a realist notion of empirical verifiability—that which would be *im*partial and *ob*jective. Yet the ways in which we imagine remote spaces are colored through and through by the ways in which we inhabit human-scale space. Such spaces, however abstracted, remain of the body, by the body, for the body. This coloring is so architectonic as to be, in certain respects, unrecognizable—a set of deeply seated assumptions about how the cosmos-as-world is. One must understand an imagined space in string theory in terms of its, as Bachelard puts it, "functions of inhabiting" (15). The "further" one is from an imagined space, the more mediations via layers of abstraction, the more fleeting the traces of the concrete seem within that imaginary. What exactly does it mean to live in a world chock-full of abstractions? There are fundamental abstractions such as objects; trajectories; events; agency; and force, and there are subtler abstractions—made subtle, in large part, through aggregation—such as heterotic superstrings and 3-branes. Each posits a path back to a provisional concretion. Each bears with it the vestigial lineaments of concretion, where concretion here means a place imaginatively trodden over and again by human feet—the things in that place touched and handled by human hands. As such, the radical heterogeneity, both in scale and kind, of a string theory imaginary is not merely arbitrary but rather, a consequence of these "functions of inhabiting."

It is in this sense that Bachelard's assertion that "all really inhabited space bears the essence of the notion of home" takes on a particularly acute significance (5). An imagined space such as that of the hadronic or heterotic string clearly is not literally made to resemble a home—with its porch, gables, carpets, coffee tables, closets, and duvets. Rather, to inhabit a space, to return to it again and again, to give it stability through repetition and redundancy, even in the imagination, is to cultivate within that imagined space the "notion" of home; it is a quasi-domestication. The objects within that imagined space gain a certain familiarity and predictability as they occupy that space and move about within it. The categorical structures that we impose on that imagined space—objects with varying degrees of parsed-out permanence, plus the rules and provisions for the interactions of those objects with each other and with the space they inhabit—grant that imagined space a homelike feel, however seemingly attenuated. Bachelard argues that a home "constitutes a body of images that give mankind proofs or illusions of stability" (17). With respect to a string theory imaginary, it is the reverse of Bachelard's assertion that proves valid: a body of images gives us proofs of stability of an imagined space. Initially, a novel space—alien, remote, inhospitable—must be imagined "very actively" in order to be experienced. This imaginative achievement allows for what Serres calls "the emergence of the object." It then becomes meaningful to speak of an object such as the heterotic string at 10^{-33} centimeters. And that object—the string—in turn, works to "stabilize our relationships" amongst ourselves and, significantly for our purposes, to that heretofore remote, alien space (*Genesis* 87).

But it would be an oversimplification to claim that an abstract space such as one manifest in a string theory imaginary constitutes a textual representation of an external reality, where a symbolic system so configured *mirrors* unambiguously an objective reality. While mathematical arguments provide constraining prompts on embodied intervention in the physical world, the abstract space of a string theory imaginary, in turn, allows for an occupation of a physical world—inaccessible directly, but nonetheless imaginable, as well as an interaction with the phenomena that populate that world. The space then, as Bachelard suggests, is "seized upon" and "lived in" with "all the partiality of the imagination"; it becomes transformed into a quasi-homelike space. Neither the mathematics nor the imaginary alone can assure positivity; they function inextricably in tandem. In this tentatively inhabited abstract space, legitimate phenomena are those things that appear of their own accord; that is, naturally. Tellingly, phenomena such as strings and branes also tend to be those things which we may conveniently imagine as experientially basic.[35]

The manipulation of abstracted mechanistic models yields to discovery of heteronomous, non-human agency, that which acts in its own right. Yet this is a process of domestication that is perpetually inconclusive, since solutions within the technical discourse generate further problems that require solving. As such, within the abstract spaces of string theory technical exposition, strings or branes are monstrous in the sense of elusive, difficult-to-grasp, natural phenomena that must be approached cautiously, obliquely, and persistently, in order to be provisionally contained (within a formal structure) and ultimately understood.

The counterintuitive quality of an imagined space in string theory might suggest what Sigmund Freud calls the uncanny. Freud defines the uncanny as "what one calls everything that was meant to remain secret and hidden and has come into the open" (132). While the notion of the uncanny engenders the affective tension between the "homely" (*heimlich*) and the "unhomely" (*unheimlich*), the uncanny pertains, in Freud's reading, to "what was once well known and had long been familiar" that only later becomes "unfamiliar" or discomfiting (124).

Yet for a string theory imaginary, the reverse is the case. Such an imaginary initially embodies a space that was never well known—a radically unfamiliar, alien, and remote place, a place that is very much, while unknown and uncertain, imagined specifically for the purpose of having it come out into the open. Paradoxically, the very act of abstraction draws such a place not away but provisionally (in the case of string theory) into "plain sight." For Freud, the uncanny represents an "unintentional return," whereas with a string theory imaginary, the epistemological exploratory voyage is very much intentional, and as such, is not a *nostos*, a homecoming (144). That which is homelike is extended outward rather than reencountered from a place imagined as abroad. While aspects of the imagined space may appear familiar, these traces borne by the imagery are imported into the space, not discovered there. The importation of the homelike with respect to the imagery invoked in string theory technical discourse affords a means of grappling abstractly with the radical alienness of this place, discovered *in media res*. The alien thus discovered is not uncanny in the sense that Freud would have it, as primitive totems once buried deep within the individual and collective unconscious, but rather, weird or monstrous in the sense of oracular. The next chapter explores, among other things, the extent to which the promise of the domestication of this alien space and the phenomena that inhabit it offer the prospect of a certain power over a more broadly construed human destiny.

It is important to reemphasize that there is no reason to expect that the physical world at the scales of 10^{-16} or 10^{-33} centimeters are "places" that are capable of being apprehended in an unmediated, direct fashion. It is an utterly and categorically alien realm. An imaginary offers an abstract space that mediates between human-scale agents and remote-scale alien phenomena. These agents take, set up, operate upon, and purify objects within the imaginary. Phenomena—strings and branes—emerge from this imagined interaction to express their heteronomous agency. The objects, understood as basic, gain substance through these imagined encounters. Ultimately, an imaginary grants theorists a critical means for doing their work all the more effectively. Its peculiar features become a convention passed on from one article to the next that forms a tradition.[36] A string theory imaginary helps to support and sustain that work, which consists principally, as theorists would argue, of mathematical formalization. It is work conducted on whiteboards, scratch pads, and computers, through conversation and the building of consensus. As such, a string theory imaginary, in many respects, bears with it traces of the emotional content of that praxis.

If one isolates the narrative structures embedded within the various imaginaries thus far examined in terms of their affective resonances, a pattern emerges. As a narrative, a string theory imaginary in technical discourse begins to resemble romance—in a heroic mode. I am, by no means, the first one to suggest this. In *Beamtimes and Lifetimes*, Sharon Traweek observes that, within the culture of high energy physics, the "textbook images of the scientist-hero, his personality and his exploits, have much in common with the fictional mode of romance" (81). She describes how practicing particle physicists tend to see the history of their discipline as "a short hagiography and a list of miracles."

> In their careers, physicists journey from romantic readings of others' lives, through handing on mimetic tales of heroic action and quests for survival, to becoming skilled practitioners of gossip and rhetoric. They complete the circle by telling erotic tales about physics, tales transformed into romance for the next generation of neophytes. (77, 103)

What Traweek refers to here are the discourses (and their concomitant narrative modes) of recruitment and pedagogy, such as textbooks, as well as the informal communications that take place at conferences or inside classrooms and laboratories. Yet one also can find clear traces of this romantic narrative mode in the technical articles examined in this chapter.[37]

The account of heroic romance that follows draws on the work of Northrop Frye, and in particular, his classic work, *Anatomy of Criticism*. While narratological analysis has been a central concern in more current work in science studies, I use Frye here not out of neglect for these recent developments, but rather, for two crucial reasons. Firstly, it is Traweek herself that cites Frye in her reading of particle physics culture as romance. By applying Frye to string theory technical discourse, I want to extend that reading and demonstrate that the heroic mode that so permeates imaginatively the communal aspects of particle physics culture—textbooks, departmental and laboratory social dynamics, informal communications, etc.— also pertains to its technical exposition. Traweek's work merits a central role because, as an ethnography, it remains seminal, both for its comprehensiveness and rigor, and in its investigation of the culture of particle physics, a culture that has remained relatively stable over the course of the past two decades. Secondly, the more recent work in narratology, whether considered generally, or as it applies specifically to science studies, is, as of yet, not especially suited to my purposes. As Monica Fludernik concedes in *A Companion to Narrative Theory*, "current narratological categories are still by and large based on the novel between (maximally) 1700 and the 1990s" (50), rather than the technical exposition of theoretical physics. More recently, there have been concerted efforts to broaden the scope of narratological analyses beyond the novel to embrace other media such as film, face-to-face conversation, folktales, and legal documents, to name a few.[38] Yet the continued emphasis on the narratological analysis of, first and foremost, imaginative writing, tends to limit the scope and applicability of the methods developed for the kind of close reading relevant to a study such as this one.

One notable example is the recent work of David Herman. While the methodology that Herman elucidates in *Story Logic*, for instance, constitutes an important contribution to the field—in particular, his explication of the role that states, events, actions, participant roles, temporalities, and spatialization play within what he calls storyworlds, *Story Logic* as a whole focuses on imaginative writing or other comparable media, rather than tackling scientific discourses directly as narrative forms. At the risk of overgeneralizing, I would say that this is the case with most of the work in what recently has come to be known as cognitive cultural studies.[39] Conversely, recent science studies work that employs narratological analytics also loses relevance insomuch as it considers, by and large, scientific disciplines at an epistemological remove from theoretical physics—for instance, ecocriticism, bioinformatics, or epidemiology.

One notable exception is Bruce Clarke's discussion of the nineteenth-century science of thermodynamics in *Energy Forms*. Clarke argues that scientific models such as thermodynamics "embody imponderable concepts in tangible forms" (17). Physical theories gain "conceptual and technological purchase on the imponderable forms and phenomena . . . not by seizing reality bare-handed but, to a significant extent, through *scientific* allegories, that is, by constructing and investigating as factual fictions increasingly workable models" (18, emphasis in original). For Clarke, allegories are "intrinsically anachronistic structures that interfold the historical past, the perceived present, and the imagined future" (6). These structures "mediate and moralize the relations of different places and eras, disparate realms of activity, and different levels of being" (17). As structures that mediate between the social and the natural world, allegories necessarily gravitate "to agencies of power, molding as well as mapping the power relations of competing discourses" (23). They function, on the one hand, by effecting "rhetorical transpositions of inanimate processes into reified personages" whereby such "reification concretizes abstractions into things, often as agents of some sort" (20, 30). On the other hand, "allegory typically creates a cosmos, a hypothetical or fictive totality" (27).

When one considers string theory as an imaginary framed as an encounter in a remote, alien space between an abstracted human agency and a natural agency in the form of strings or branes, Clarke's notion of scientific allegory becomes entirely relevant. Yet in *Energy Forms,* Clarke is concerned principally with the popular dissemination of thermodynamics (as well as the ether and the fourth dimension), not the imaginative realization of the science of thermodynamics within its technical discourse. The term scientific imaginary is, to a certain extent, a more precise formulation of "scientific allegory" insomuch as it accommodates a degree of abstraction beyond the more readily recognizable allegorical forms present in self-avowedly imaginative writing—or even popular science. It is also worth pointing out the overlap between Clarke's notion of scientific allegory and Frye's take on romance. In effect, romance engenders a specific kind of allegorical reading particularly relevant to string theory technical discourse.

Let us then define more precisely romance as a narrative mode. Its basic form follows the following arc: an idealistic hero patronized by some higher authority undertakes a quest or journey to places remote from ordinary life in order to complete a task and/or return with an object desired by that authority. In *Anatomy of Criticism,* Frye explains that a romantic narrative "has three main stages: the stage of the perilous journey and the preliminary minor adventures; the crucial struggle, usually of some kind of battle in which either the hero or his foe, or both, must die; and the exultation of the hero" (187). Frye observes that:

> The essential element of plot in romance is adventure, which means that romance is naturally a sequential and processional form . . . At its most naïve it is an endless form in which a central character who never develops or ages and goes through one adventure after another. (186)

Frequently the hero's quest involves a journey to a "lower world" or a wilderness, "a place of oracles and secrets" (193). The object of the hero's quest typically stands for "wealth in its ideal forms, power and wisdom"

(193). The servants and friends of the hero "impart the mysterious rapport with nature that so often marks the central figure of romance" (197). Frye states:

> The characters who elude the moral antithesis of heroism and villainy generally are or suggest spirits of nature. They represent partly the moral neutrality of the intermediate world of nature and partly a world of mystery which is glimpsed but never seen, and which retreats when approached. (196)

A moment of epiphany marks the point in the narrative arc where the hero's agency comes into alignment with the world of nature.

The articles considered in this chapter share, to a certain extent, these same basic structural elements of romance. Firstly, there is a narrator cloaked in the formal self-abnegation of the first person plural.[40] The narrator is idealistic in that, by undertaking a given enterprise, he or she believes in the pursuit of knowledge ("wisdom") by these means. Optimism—implicit in the conceit that the article identifies a worthwhile problem and successfully offers a solution to that problem—informs the narrative tone. Furthermore, the narrator recognizes and identifies with a community of peers—those working on the same problem and those charged with evaluating the merit of the article's technical content. The framing structure of the exposition resembles the form of a quest: a problem is delineated, permutations are explored, and a solution is offered. As such, technical exposition constitutes what Frye calls "a sequential and processional form." Problems lead to solutions, which in turn lead to further problems and solutions without any sense of finitude or arrival. The individual articles link up to this perpetual chain of progress by anticipating further work: Susskind is concerned with "the problem of removal of unphysical states" (494), Gross et al. with "deriving all of known physics from the . . . heterotic string" (283), and Randall and Sundrum with "solving unresolved issues in quantum gravity and cosmology" (8). The work of theoretical physics is, in this sense, endless.

Strictly speaking, the narrators of these texts never develop or age. The arguments within the articles develop, but such progress reflects solely on its subject-matter—on theory or model standing metonymically for the cosmos. Accordingly, it is not the narrators that age but the scientific knowledge that they produce. Theorists must constantly contend with the rapid obsolescence of their work, with, according to Traweek, its roughly six-month life expectancy. The narrators belong to and participate in what she calls "an extreme culture of objectivity": within the imaginary, they would have their voices purified of personality, gender, historicity, contingency, embodiment (162). They exist within a timeless space of abstraction (the space itself contains time) where theoretical objects are held up to abstracted scrutiny and manipulation.

As in romance, the spaces within these imaginaries are initially situated far off from the familiar territory of ordinary life. For instance, hadronic string theory explores the inner mysteries of the atomic nucleus—a scale removed from human gazing and grasping by sixteen zeroes (10^{-16} centimeters). This is a place inhabited by exotic objects such as quarks, mesons, five-dimensional harmonic oscillators, strings, and "ghosts" ("unphysical" states).

Problem-solving is framed in the following terms. Narrator(s) work with the mathematics—couched by the supporting exposition in terms of composing, taking, setting up, building up, etc.—in such a way as to allow nature to reveal itself, to exhibit what previously was secreted and secretive "behavior." It is in this sense that strings (and branes) serve metonymically for the cosmos. They are agents—neither purely inert objects nor autopoietic organisms, but, depending on the imagined context, exhibiting characteristics of both. Consequently, these phenomena—strings and branes—possess a certain "moral neutrality," indifferent to the rivalries of the various groups of theorists. They must be approached obliquely, cautiously, and rigorously, lest they retreat into obscurity. Once they do reveal themselves, they are then, in the imaginary, understood. To imaginatively stand among them is to give them substance. Being understood, they then can be brought back out of the unknown—a highly abstracted form of wilderness, analogous to Frye's concept of the "lower world"—to be offered up to patronizing authorities as a prize, a quasi-domesticated object. In effect, within the romantic mode, the abstract space of the imaginary becomes what Frye calls an "intermediate world of nature."

In each imaginary, the narrative tension builds up to moments of epiphany, when the narrator's agency comes into alignment with nature-as-it-is. In delineating the features of a particular configuration of the heterotic string, Gross et al. declare that "already at this level there are 18 883 584 physical degrees of freedom!" (266). Randall and Sundrum write of the "*dramatic* consequence" that "we can live in 4 + n non-compact dimensions" and, in another instance, the special "state" that concerns them "is *indeed* a bound state" (1, 2, emphasis my own). The moments of epiphany in these examples have been pared down to an exclamation point, to the enthusiastic excess of the inserted qualifiers *dramatic* and *indeed.* A pivotal moment of this type also occurs for Susskind in section ten of "Dual-Symmetric Theory."

> Le [*sic*] us suppose that a meson is composed of a quark-antiquark pair at the ends of an elastic string . . . As the string moves in space-time a two-dimensional strip bounded by the trajectories of the quarks is generated. (483–84)[41]

Here the work of mathematics has been accomplished meticulously enough such that a meson divulges its inner mystery. Susskind promises the reader that all he or she has to do is the work of supposing. Out of this modest effort, a string arises naturally. It moves and from that movement something marvelous and potentially valuable is "generated" (a word that bears with it a trace of the sense of birth)—the world-sheet. This stepping-aside, so to speak, of the narrator at this moment marks the kind of effacement that occurs in romance; it is the ritual-death of the romantic hero in a highly abstract form. Out of this effacement arises the possibility for the narrator to discover nature, rather than merely to invent some speculative "unphysical" model. Implicit in the tangible evidence of the article's publication—something that every reader is cognizant of—is the reward for such effacement. Having survived peer review, the narrator may now be duly recognized for his accomplishment, a genuine discovery.

Traweek speaks of the "denial of human agency in the construction of science" (158). As I have argued, that agency is not denied in this romantic mode, but rather, abstracted in an imaginary, drawn away from its human origins but nevertheless retaining traces of that humanness. The romance that takes place within string theory technical exposition allows the theorists to, as Le Doeuff puts it, "sustain something which the system cannot itself justify, but which is nevertheless needed for its proper working" (3). More than merely affording a ludic or didactic context of discovery for their work, a romantic imaginary offers string theorists a means of: grounding a highly abstracted practice within a broader cultural tradition—namely, an admixture of quasi-domesticating and wilderness-like exploration frames of imaginative reference; substantiating and thereby justifying theoretical objects (and spaces) that otherwise would be entirely inaccessible and incoherent because, as Le Doeuff contends, they possess a "meaning . . . incompatible with the system's possibilities," the system here being the "insubstantial" formalisms of mathematics (3); and providing a means of charging esoteric practices with a highly motivating and glorifying affect. All three means speak directly to the Durkheim supposition: the romance of imaginary-as-cosmos blends with the romance of imagined social praxis to affect a way for theorists to imagine themselves into a structured and justifiable relationship with the cosmos.

CHAPTER 4

Accessibility and Authority

String Theory Popularizations

Pedagogic Space of Access

The previous chapter surveyed the expository portion of certain representative texts in string theory technical discourse in their development of a scientific imaginary. The content of this exposition is best understood as an imaginary because it is primarily concerned with substantiating an abstract theoretical space as that which is natural. Theorists attempt to expose this abstract theoretical space as something physically coherent by means of a procedural space that intersects with the theoretical space. At the nexus of theoretical and procedural spaces, what the theorists construct gives way to an encounter with natural phenomena in the form of the strings and branes that populate those spaces. In the logic of the imaginary, they then can claim, assuming sufficient mathematical rigor, to have made a genuine discovery of the nature of the cosmos.

As an imaginary, the abstract theoretical space consists of a complex aggregate of recombinant human-scale images, and as such, possesses a radical heterogeneity that undermines a coherent "physical interpretation" of the phenomena exposed in juxtaposition to mathematical arguments. Nevertheless, as a place of encounter between theorists' procedure and natural phenomena, the imaginary within technical discourse substantiates those phenomena by affording a quasi-domestication of that space—of rendering the remotely alien in familiar human-scale terms. This quasi-domestication works to counteract the incoherence of heterogeneity. In some respects, those natural phenomena—strings and branes—become a matter of imaginative convenience because they are, within the context of the heterogeneous theoretical space, couched in cognitively basic terms.

Furthermore, an epistemological regime that, rather than being conceptually "pure," incorporates a certain cultural conditioning of affect, helps to sustain this imagined encounter between the human and the natural. As a consequence, the alien microcosm is made all the more familiar and thus knowable—an ideal confluence of that which is both manipulable and observable. The key dynamic within the imaginary of technical exposition, then, is that of an encounter within an abstract theoretical space.

This chapter assesses the imaginaries within another, derivative yet distinct discourse—string theory popularizations. For popularizations the crucial dynamic of encounter in technical exposition gives way to that of *access*—the presentation by an authority of a modified version of the theoretical space of string theory to a non-specialist audience. I have chosen the term *access* for two reasons. Firstly, it is a term string theory popularizers themselves use: a purpose, invariably professed within the text, for writing the popularization, and a justification for the contrast in expository content to that of technical discourse. By theorists' own admission, different audiences necessitate a different treatment of the material. Secondly, the notion of access complements the inferential structure implicit in string theory imagined as an abstract theoretical space—such a space being first and foremost a place in which one may imagine objects in fixed positions or moving about—a space that one may imagine oneself coming into or approaching. Moreover, since it is a remote, alien space, it follows that it is not so easily approached. Therefore, in the logic of the imaginary, the uninitiated require a guide in order to access the space successfully.

This imaginative frame of a remote space through which the non-specialist reader may gain access exclusively by means of a guide conditions a particular relationship between the implied author of a popularization and his or her implied readership. It becomes, in effect, a monologue delivered by the narrator as an authority to his or her audience. In "Between Fact and Fiction," Felicity Mellor argues that, with respect to science popularizations, "authors are made present and relevant only through the reading of the texts" such that the "audience will construct an understanding of the author out of the clues within the text and the web of texts surrounding it." As a consequence, "the author of a popular science book is significant as a textual presence"—as an "implied author" (519). In keeping with Mellor's formulation, I will refer to the authors of string theory popularizations in this sense, as an implied *textual presence*. This implied textual presence, as the narrative voice of exposition within the popularization, may claim for itself the role of guide within the imaginative frame of access insomuch as it establishes its own authority with respect to the abstract theoretical space of string theory. Given the importance of guided access—by an implied author, duly authorized, on behalf of an implied audience—this chapter addresses the following question through the close reading of selected popularizations: precisely what is at stake when theorists make the abstract theoretical space of a string theory imaginary accessible? As with the previous chapter, a close reading of key passages in the texts is required in order to examine thoroughly the specific imaginative and epistemological contexts in which core images are situated.

The three popularizations I have chosen for close reading are Michio Kaku's *Hyperspace: A Scientific Odyssey Through Parallel Universes, Time Warps, and the 10th Dimension*, Brian Greene's *The Elegant Universe: Superstrings, Hidden Dimensions, and the Quest for the Ultimate Theory*, and Lisa Randall's *Warped Passages: Unraveling the Mysteries of the Universe's Hidden Dimensions*. These three texts are representative of the entire genre in large part because they each meet several criteria. Firstly, the author is a practicing and accomplished theorist, which serves to validate authority. Secondly, when it was published, the book was timely and commercially successful, thereby assuring a relatively wide audience. And lastly, the text generally exhibits a balance between originality and conformity to the imaginative conventions of the genre. This last criterion speaks directly to the issue of access: with respect to an imaginary, commonalities in the various presentations will serve to highlight those genre conventions.

The first popularization under consideration, *Hyperspace*, was published by Oxford University Press in 1994. While not the first string theory popularization, *Hyperspace* enjoyed significantly more critical attention than its predecessors and was a great deal more commercially successful.[1] One can attribute the increased exposure that *Hyperspace* received to both timing and follow-up momentum. The success of preceding texts such as *A Brief History of Time* by Stephen Hawking served to expand public awareness of theoretical physics. *Hyperspace* arrived in the marketplace during what many consider to be a boom in popular science publishing, which occurred in the mid-1990s. The author of *Hyperspace*, Michio Kaku, is a prominent member of the string theory professional community. For the past twenty-five years, he has held the Henry Semat Chair and Professorship in theoretical physics at the City University of New York (CUNY). He made his reputation in the field in the mid-1970s with his work on string field theory—the first rigorous synthesis of string theory with quantum field theory.[2] Kaku has become a preeminent spokesperson for string theory to the public. Since *Hyperspace*, he has continued regularly to publish popularizations.[3]

Brian Greene's *The Elegant Universe* arguably is the most successful and widely read string theory popularization to date. First published in the UK by Jonathan Cape and in the U.S. by Random House in 1999, it went on to become an international best seller and won the Aventis Science Book Prize for 2000.[4] The science program *NOVA* also adapted the book into a highly successful three-part documentary, which Brian Greene hosted, for the U.S. Public Broadcasting System (PBS). Greene made his reputation as a postdoctoral fellow at Harvard University in 1987 with his contributions, along with Ronen Plesser, to the subfield of "mirror manifolds," a term he and Plesser coined.[5] This reputation was further cemented with his work in 1992, along with David Morrison, Paul Aspinwall, and Ed Witten, on "space-tearing flop transitions."[6] Currently Brian Greene is a Professor of Mathematics and Physics at Columbia University.

HarperCollins published Lisa Randall's *Warped Passages* in 2005. Randall is a Professor of Physics at Harvard University. As mentioned in the previous chapter, she does not categorize herself as a string theorist, but rather, as a model builder.[7] She made her reputation principally from three widely circulated papers: one, coauthored with colleague Raman Sundrum in 1998, on "sequestered supersymmetry breaking" in a braneworld scenario (*Warped* 334–50); another, published in 1999, proposed a solution to the hierarchy problem by means of a braneworld scenario that involves warped geometry; and "An Alternative to Compactification," which the previous chapter examined in detail.[8] *Warped Passages* was a *New York Times* "Notable Book of the Year." The dust jacket biography states that Randall "was the first tenured woman in the Princeton physics department and the first tenured woman theoretical physicist at MIT and Harvard. Her work has attracted enormous interest and is among the most cited in all of science."[9] I have chosen *Warped Passages* because it exemplifies the latest popular treatments of current trends in research following the 1995 "second superstring revolution."[10]

In her monograph, *Reading Popular Physics*, Elizabeth Leane observes that the "popularization of science has traditionally been viewed . . . as a flow of established, unalterable knowledge from an authoritarian scientific community into a passive public" (9). That this scientific community serves and ultimately depends financially on the public does not preclude most scientists—and those who publish popularizations—from assuming that a lay audience is not in a position to comprehend, let alone appreciate, the rigor of a technical discourse such as that of string theory.[11] This image of outward flow from a concentrated enclave to a wider field often implies a dilution—if not contamination—of scientific knowledge, which betrays a certain condescension on the part of the disseminator. Consistent with this attitude, it is no surprise that the term *popularization* often comes with derogatory connotations, and that many consider popularizations a dumbing down of science in keeping with a least common denominator populism that attempts first and foremost to secure the widest possible market. In light of this widespread assumption, popularizers keen to position their texts advantageously in the marketplace emphasize their text's pedagogic value, primarily by hewing closely to what Stephen Hilgartner calls "appropriate simplification," as opposed to mere "distortion" (519).

The previous chapter began with the simple observation that string theory technical discourse consists of the following principal components: mathematical argument and an expository imaginary. The imaginary instantiates an encounter by a procedural space—the doing of string theory—with an abstract theoretical space—the putative results of theorizing, taken ultimately as the cosmic order. The popularizations under consideration, in keeping with convention, omit both mathematical argument and that procedural space. As discussed in chapter 2, in a realist stance, it is mathematics that constitutes the entire truth-value of technical discourse, its concept, as opposed to image. With popularizations, the non-technical content, the imaginary of technical exposition, paradoxically, is adapted to serve as the conceptual content of the text. Mathematical argument, heretofore the *de facto* conceptual content of technical discourse, becomes relegated to, at best, endnotes—if not completely excluded from the text, a fleeting presence at the margins of the text.[12]

Given a shift in emphasis within the imaginary of these popularizations from encounter to access, this chapter explores the extent to which the abstract theoretical spaces undergo a transformation such that they engender an alternative epistemological imaginary to that of technical exposition. In chapter 2, I explored some ramifications of the fact that a great deal of one's commonsense understanding of the world is a form of received knowledge; that is, knowledge consumed secondhand by means of an imaginary endorsed by an authority. As such, the legitimacy of an epistemological imaginary is principally a faith-based affair: I believe in heliocentrism not because I have conducted the experiments myself to prove its veracity, but because others supposedly have, and I take their word for it. This is not to say that heliocentrism is not an empirically legitimate theory, but that for those who are not physicists, the physics of heliocentrism will be no more than an epistemological imaginary. As an imaginary, its reality is substantiated not by experimental validation but by social convention—by the cultural currency of its constituting images and conceits, and by the ways a given community's members embrace the social practices associated with that imaginary. Believing in heliocentrism, I behave differently toward what I understand the cosmos to be—and toward others within my community—than I would were I to subscribe solely to geocentrism. In effect, heliocentrism as an imaginative conceit has values attached to it—social attitudes and norms—that differ from those of geocentrism. Accordingly, it is my commitment to social practices predicated on those attitudes and norms that then further substantiates heliocentrism, rather than my conducting an actual experiment to verify its reality.

As discussed previously, Michel Serres calls attention to the tendency for scientific discourse in general to eschew external philosophy and arrogate to itself an *endo*-epistemology. It is specifically an *endo*-epistemology because its imaginative regime is presented in such a way that it is taken as endogenous to the content of the exposition itself, as both meta-discursive envelope and integral internal constituent. For example, within string theory technical exposition, the heterotic superstring is imagined both as substantial phenomenon within a heterogeneous abstract theoretical space and, within the epistemological imaginary, as that which lies beneath a surface. The heterotic superstring has substance as the "deeper" reality. Ultimately, this chapter examines how the pedagogic spaces presented in the selected popularizations incorporate an epistemological imaginary that legitimates the string and brane of string theory in the face of other competing or preceding theoretical objects.

Kaku's *Hyperspace*

About halfway through *Hyperspace*, Kaku preludes an extended explication of string theory's principal features by declaring that if string theory is to supersede quantum theory, the string itself must be afforded a privileged status as a physical phenomenon over its predecessor—what he calls, "force."

> If our scientists invent concepts like forces, it is only because they cannot visualize the invisible vibrations that fill the empty space around us. Some scientists sneer at the mention of higher dimensions because they cannot be conveniently measured in the laboratory. (5)

There are two images set up in competition here. One posits force as an ostensibly "visible" phenomenon that plays out within the "empty space around us." Force is imagined in an object-event frame where the notion of force is assigned to the imagined effect of an action—the interaction of matter as particle-object(s) with itself.[13] The other frame imagines those same phenomena as "invisible vibrations that fill" empty space—a location-event schema where the notion of force corresponds to the phenomenon of space-time itself, rather than to the effect of particles-as-objects interacting. This alternative imaginative frame allows one, paradoxically, to "visualize" these "invisible" vibrations—that which before was only indirectly observable becomes imaginable, for to "visualize" is, strictly speaking, not to see directly but to imagine seeing. Matter, as form, and force, as the effect of the actions of that form, turn out to consist of the same substance—that which vibrates, hyperspace. Kaku then complains that realist scientists "sneer" at this more fundamental phenomenon, because it is "inconvenient" to measure higher dimensions in the laboratory—a setting that presumably reinforces the overemphasis—in his epistemological imaginary, on a superficial failure of visualization, one dependent on measurement. In this example, Kaku presents the abstract theoretical space of string theory (vibrating hyperspace) in tight conjunction with an epistemology (it is an invisible, yet visualizable, substance) that justifies its reality, principally by means of his pedagogic authority.

Kaku later frames this problem of the inadequacy of the concept of force in terms of "puzzling questions" that require answers—answers that string theory provides.

> The deeper we probe into the nature of subatomic particles, the more particles we find. The current "zoo" of subatomic particles numbers seven hundred, and their properties fill entire volumes. Even with the Standard Model, we are left with a bewildering number of "elementary particles." String theory answers this question because the string, about 100 billion billion times smaller than a proton, is vibrating; each mode of vibration represents a distinct resonance or particle. The string is so incredibly tiny that, from a distance, a resonance of a string and a particle are indistinguishable. Only when we somehow magnify the particle can we see that it is not a point at all, but a mode of vibrating string. (152–53)

According to Kaku, the puzzle-solving process of string theory practice is a form of "prob[ing] into the nature of subatomic particles." While *to probe* has a more generally abstract denotation of "to test," Kaku's use here exploits the term's connotation of an exploratory action with a small instrument that is directed inward and downward into a remote space (as opposed to an outer space probe that goes upward and outward). This probing is imagined through a "we" that travels "deeper . . . into the nature of subatomic particles." In this imaginary, the multitudinous superfluity of "particles" hides a deeper reality where those particles are to that underlying nature as form is to substance. As "we probe" more deeply, "we find" ever more particles—a conceit that, in some respects, reimagines the microcosm in terms of the macrocosm. Like the stars, nebulae, and galaxies of outer space, these microscopic particles are, on initial impression, "bewildering." Their chaotic profusion provokes confusion. But this ferocity of profusion becomes tamed as "entire volumes" are filled with their "properties," an image that plays on the Galilean conceit of nature as text, as that which is decipherable and ultimately legible (237–38). Being legible, within Kaku's imaginary, quanta have been thoroughly, if not coherently, subjected to a kind of taxonomy. Physicists have distinguished and ordered them such that, within the logic of the imaginary, quanta now comprise a "zoo." This image of the "zoo" of subatomic particles in quantum theory occurs with such frequency in popular accounts of contemporary high energy physics that one may justifiably consider it a stock convention. It evocatively captures the dynamic tension between the "wildness" of subatomic particles—the sheer variety of attributes and interactive behavior—and the achievement of a containment or quasi-domestication of them that quantum theory represents. In effect, completely in keeping with convention, Kaku initially blends three relatively contradictory imaginative conceits to make a case for the string: quanta as elemental form, as animal-like (at least in one respect), and as text-like.

Enveloping this triad is a fourth epistemological conceit: the "question" of the subatomic "zoo" is something for which string theory provides an answer—in this instance, through the image of the string. Playing on the imaginative conceit of quanta as superficial form, Kaku posits that the string resides still deeper within subatomic space, functioning metonymically as the cosmos. He then offers a comparison between one quantum particle, the proton, and the string, which, he writes, is "about 100 billion billion times smaller." While in other image schemas, quanta such as protons might be marked as that which are incredibly tiny, here they are relatively huge—a billion billion times larger than the string. This is a shift in scale that the imagination is utterly incapable of comprehending. The evocation of that which exceeds the imagination yields to an appeal to affect: Kaku qualifies the string as "incredibly" tiny. One may then imagine "tiny" on the remotest of human scales, for example, as a speck of dust or a pinpoint, and then extrapolate that image beyond the ultra-miniscule into a space marked as "incredible." This tiny string is literally unbelievable and thus, for its imagined substantiation, demands a quasi-religious leap of faith to offset its lack of comprehensibility. As the authority, Kaku expects that a passage beneath and beyond that which is credible or commonsensical to the *in*credible requires a specific emotional response—specifically, astonishment or awe. The appropriate simplification of access engenders a close coupling of affect and epistemological justification at this imaginative intersection of pedagogic and abstract theoretical space.

These strings, which are incredibly tiny, "vibrate." Kaku explains that "each mode of vibration represents a distinct resonance or particle." This equivalence between the point-particle proton, and the tinier one-dimensional vibrating string suggests two paradoxes within the imaginary. The first lies in the notion of vibration. *To vibrate* is simply to move rapidly to and fro, to shake. On human scales, vibration is an event to which one may bear witness—it is observable as, for example, the vibrations of a bee's wings. Conversely, vibration is an experience that is more readily felt when the motion to and fro begins to exceed our capacity to see, as in the case of the vibrating of a mobile phone or mild earthquake. At the cellular scale, all five senses are fundamentally haptic—the touching of particles on the variously attuned cellular surfaces, whether the eardrum, skin, retina, mucus membranes, etc. When the motion to and fro of vibration is too rapid to be seen or felt—for example, the vibration of sound on the eardrum, or at the perceptual extreme, the vibration of light as it hits the retina—vibration can only be imagined as an idealized motion to and fro, such as the "waves" of photons that constitute light. Therefore, to imagine a string—a billion billion times smaller than a photon—vibrating is to render it accessible. Much like heliocentrism, as an abstraction; an event neither "directly" seen nor felt, the image of vibration becomes an epistemological conceit—an idea known intimately through the mediations of imagery drawn from human scales.

In the spirit of appropriate simplification, Kaku decides to qualify the string as something that vibrates—rather than, for instance, as something that oscillates. The term *oscillation* is synonymous with *vibration*, but retains a more abstract or technical connotation; it at least gestures toward the imaginatively remote. To write then of a vibration on the Planck scale is to locate the action that the term signifies at a scale where the distinction between abstraction and concretion loses relevance. Yet to do this, the reader must "somehow magnify" the string to human scales, placing it firmly within the abstract space of a particular imaginary. To magnify the string, then, the seeing/feeling imaginative frame of reference must shrink or the string must be monstrously inflated to human scales.

The second paradox in this image rests with the contradiction of having to imagine simultaneously a proton as point-particle and a string as one-dimensional vibrating object and then juxtapose the two in the mind's eye. "From a distance," Kaku asserts, "a resonance of a string and a particle are indistinguishable." While the "distance" to which Kaku refers, as a vantage point, is highly ambiguous, the reader nonetheless is asked implicitly to shift between two scales, the scale in which an object is point-like, and a scale, the Planck scale, in which that same object is string-like. In seeing *in*to (insight) that which lies beneath the surface, where probing deeper penetrates the veil of appearance and makes visible a deeper reality, the reader here is meant to privilege the "deeper" image of the string over the superficial image of the particle. In doing this, that which would seem to express, on first impression, the very limits of diminution—the point, reveals another layer of reality with supplemental internal structure—the vibrating string. The paradox lies in the synchronic schism between these two imagined spaces. One must imagine a journey from the one scaled space to the next, the one that lies deeper below it, in order to be able to "see" the object transformed from particle to string. Strictly speaking, a point is an object without dimension; as one approaches, a point would always remain a point. But here the point expands as one approaches it. One then enters into or passes through this exploded object to another space, at the center of which one gains access to a vibrating string. This imagined journey privileges this "deeper" abstract theoretical space as reality over the "appearance" of the preceding/surface space of the point-particle.

Concomitantly, this journey deeper justifies the status of string theory itself. There is a slippage between the string as image within the abstract theoretical space of string theory and string theory itself as an image within the enveloping pedagogic space. Kaku, perhaps inadvertently, even makes this conflation explicit: "Today, this bewildering collection of subatomic particles can be explained as mere vibrations of the hyperspace theory" (16). Here it is not hyperspace itself that vibrates, but hyperspace theory. In the imaginary, "going deeper into" becomes synonymous with "can be explained as" such that the epistemological transforms into the objectively natural.

Comparing Kaku's image of the string with the presentation of the heterotic superstring in technical discourse, one notices crucial differences. But I hesitate to suggest that the image of the string in *Hyperspace* merely represents an oversimplification compared to the heterotic superstring's intricate heterogeneity. Kaku's string also manifests a certain heterogeneity—in its admixture of string-as-object and string-as-location (where hyperspace vibrations generate quanta) and in its vacillation between quantum and Planck scales, between images of points and spaces. Rather, any given complexity stemming from a technical imaginary's need to account for the subtleties of mathematical argument becomes displaced in *Hyperspace* into the complexities of an epistemological imaginary. Kaku strives to contextualize the image of the string in terms of several enveloping imaginative frames: cosmos as mystery where theorizing is puzzle-solving or the answering of a question, cosmos as wilderness-like where theorizing is a quasi-domestication, cosmos as legible text where theorizing is reading, and finally, cosmos as the deeper reality where theorizing is a journey down through a surface. In the technical article on the heterotic superstring examined in the previous chapter, such epistemological concerns only come to the fore in a largely speculative conclusion. In *Hyperspace*, epistemology explicitly infiltrates the abstract theoretical space throughout, since such framings are duly permitted by pedagogic authority.

Having introduced the principle features of the string and the means by which it reconciles quantum theory and general relativity, Kaku endeavors to address a deceptively simple question—"Why strings? Why not vibrating solids or blobs" (156)? In the explication that follows, Kaku draws comparisons between the strings of string theory and string-like objects in other scientific disciplines, notably, biology.

> A moment's thought, however, will reveal that nature has reserved the string for a special role, as a basic building block for other forms. For example, the essential feature of life on earth is the stringlike DNA molecule, which contains the complex information and coding of life itself. When building the stuff of life, as well as subatomic matter, strings seem to be the perfect answer. In both cases, we want to pack a large amount of information into a relatively simple, reproducible structure. The distinguishing feature of a string is that it is one of the most compact ways of storing vast amounts of data in a way in which information can be replicated. (156)

Here the imaginary expands to incorporate that which will add to the legitimacy of the string as fundamental object. Again, this discussion is framed within the epistemological conceit of appearance as surface hiding a deeper reality. Also, Kaku evokes another stock imaginative conceit wherein life is understood in terms of theatre. Nature, as director, "has reserved the string for a special role." In the drama that is the universe, the string becomes the principle actor that drives the action of the plot. In a quick shift of imaginative frame, Kaku then states the string's role is as a "building block for other forms." This imaginative frame derives from the stock conceit of nature-as-building, where the part, string-as-block, ultimately defines the whole, namely, cosmos-as-edifice. These two conceits, cosmos-as-theatre and cosmos-as-edifice, enjoy a great deal of cultural currency such that they work further to substantiate the argued-for reality of strings.

Kaku then draws a connection between the string of string theory and "the stringlike DNA molecule," which he calls "the essential feature of life on earth." Presumably, the string's legitimacy as "essential feature" of nature bears a striking resemblance to the "stringlike" DNA's essentiality. Kaku effectively borrows the wide cultural currency of a particular popular discourse on DNA, with all its substantiality, to corroborate the legitimacy of strings in string theory. This popular DNA imaginary hinges on the stock conceit of code. Just as DNA encodes the information essential to life, strings encode the information essential to matter and force, to that which comprises the universe. I want to stress that the prominence of DNA in the popular imagination concerning life's essence has much to do with the popularizing efforts of the likes of Richard Dawkins, among others, who exploits the code or life-as-text conceit. Many biologists would contend that DNA is not the essential feature of life. Many primitive life forms do not have DNA; for example, certain bacteria. At the very least, some would argue that a metabolism, RNA, and/or the cell wall are equally essential to life. Kaku borrows not one truth per se, to reinforce another, but rather one dominant imaginary to corroborate another.

The "building blocks" of the universe become containers in which "we" may "pack a large amount of information into a relatively simple, reproducible structure."[14] Accordingly, in this construction conceit, just as a quasi-anthropomorphic "nature" builds life, by means of the information containers of DNA, string theory engineers the matter and force of the universe by means of the information containers that are strings. There is a productive intermingling of conventional imaginative conceits of theatre ("the special role" reserved for the string), construction, and informatics. This imaginary, taken in sum, is radically heterogeneous—blending as it does frames of code, container, building, etc.—in a way decidedly more familiar and quotidian than, by comparison, the relatively alien abstract theoretical spaces of technical exposition. An intensification of quasi-domestication by means of a further embellished "common-sense" allows

Kaku to refashion encounter into access. The imaginary affords the invagination of the microcosm—in this case, the subatomic realm on two distinct scales—and the macrocosm back into an abstract theoretical space where the remotely alien is shaped into a familiar, worldly space, predicated on several stock imaginative conventions. Within the imaginary, one finds an *ad hoc* congeries of scales, spaces, objects that serve to substantiate the legitimacy of the string as objective phenomenon.

Greene's *The Elegant Universe*

In the preface to *The Elegant Universe*, Greene places the abstract theoretical space of string theory into an epistemological frame strikingly similar to that of Kaku.

> Now, cutting-edge research has integrated [Einstein's] discoveries into a quantum universe with numerous hidden dimensions coiled into the fabric of the cosmos—dimensions whose lavishly entwined geometry may well hold the key to some of the most profound questions ever posed. Although some of these concepts are subtle, we will see that they can be grasped through down-to-earth analogies. And when these ideas are understood, they provide a startling and revolutionary perspective on the universe. (x–xi)

The "insight" that Greene makes accessible is into the quantum universe, presented as an image of an abstract theoretical space at scales "far below" our everyday experience, where smaller is oriented downward. He then likens the materials that make up this diminutive space to fabric. "Coiled" within this fabric are "numerous hidden dimensions." The reader is prompted to imagine, in quick succession, a tiny space, fabric, presumably made of some sort of abstracted thread-like material, and concomitantly, "hidden dimensions" that are "coiled"—inferring that these dimensions are rope-like—or perhaps thread-like in that they are also "lavishly entwined," evoking an image of embroidery. This image of an embroidered knot or coiled rope tucked within the hidden nooks of a minutely scaled "cosmic" fabric is jumbled with an anthropomorphism in that *it*, or by implication, *we* through *it*, "hold[s] the key." This "key" resolves "some of the most profound questions ever posed."

As with an imaginary in technical exposition, there is something distinctly Escher-esque about the radical juxtaposition, the imbrication, of these various and sundry quotidian objects and their shifting scales. Even Greene's use of the adjective *subtle* reinforces the imaginary of cosmic fabric in that *subtle* originates etymologically in the sense of "finely woven." This image of fabric that is both visual and tactile enables imaginative access to both the concepts and the material objects for which they double. Within the heterogeneous imaginary, an epistemology conflates string theory itself with the extra dimensions within it; both the concepts and the cosmos it describes are subtle. The sum effect of this imaginary speaks against coherence. As with Kaku, an unresolved imaginative incoherence requires, then, a framing—access as "appropriate simplification"—that would seek a certain domestication whose *fait accompli* ultimately depends upon an uncontested submission to the authoritative pedagogic voice. While ostensibly "simple" in relation to technical exposition, this is an access to an *ex*cess contingent upon a twofold straightforward acceptance: firstly, of an author's declaration that the incoherent is, in fact, coherent, and secondly, of the legitimacy of a conventional epistemological imaginary that substantiates the abstract theoretical space. Ironically, in this context, the author gains a significant chunk of his authority precisely by subscribing to these stock epistemological conventions, which enjoy a great deal of cultural currency.

Greene unequivocally states his rhetorical strategy: to present his "subtle" subject in a way that "can be grasped through down-to-earth analogies." The reader is given access, is invited to grasp imaginatively the concept-objects at this abstract theoretical height, by means of earthbound analogues—a pedagogic technique akin to that of Kaku. Considered in isolation from the subject-matter's textual source, technical discourse, such an admission plays into the realist conviction that neatly sorts the context of discovery from that of justification. Liberated from the burden of expressing the truth-value of a theoretical concept, the popularization may freely deploy imagery for purely illustrative purposes—purposes of access for an uninitiated audience. In the binary of concept and image delineated by Le Doeuff, such imagery is ludic, in contrast to the "serious" conceptual work of mathematical argument. Yet Greene is nevertheless concerned with communicating conceptual content—a content that may be subtle, but that is not impenetrable, in the way that he assumes the mathematics of string theory would be to his readers. Greene's pledge once again brings to the forefront the contradictory notion of offering access to that which is inaccessible, insomuch as it is almost entirely absent from the text. This paradox is resolved by decree: one must simply trust or, in other words, submit to the connection Greene makes between the imaginary he presents and the truth of scientific argument.

In a programmatically florid style, Greene introduces string theory to the reader.

> For three decades, Einstein sought a unified theory of physics, one that would interweave all of nature's forces and material constituents within a single theoretical tapestry. He failed. Now, at the dawn of the new millennium, proponents of string theory claim that the threads of this elusive unified tapestry have finally been revealed. String theory has the potential to show that all of the wondrous happenings of the universe—from the frantic dance of subatomic quarks to the stately waltz of orbiting binary stars, from the primordial fireball of the big bang to the majestic swirl of heavenly galaxies—are reflections of one grand physical principle, one master equation. (4–5)

Again, the image of a thread woven into a tapestry intertwines string theory in the epistemological imaginary with the string of the abstract theoretical space itself. This image of string theory as tapestry, in turn, becomes interwoven with the hagiographic image of Einstein, the theorist-hero, seeking what Greene shortly thereafter calls, resonant of Kaku, the "Holy Grail of modern physics" (15).[15] With millenarian overtones, Greene exults, "at the dawn of the millennium," the universe's "elusive" secrets "have finally been revealed." This is, once again, the stock epistemological conceit according to which the cosmos possesses secrets that lie beneath a surface of appearance. This surface/depth disparity prompts a conventional allotropy completely in keeping with Kaku: the chaotic yet "wondrous happenings of the universe," Greene promises us, "are reflections of one grand physical principle, one master equation."

Within this imaginative mix, there is something vaguely zoologic in the way Greene describes the objects that constitute the universe, using, as he does, adjectives that could have been imported from a BBC nature documentary chock-full of lavish wide-angle IMAX shots of African savannahs and narrated by Sir David Attenborough. There is a "frantic dance," a "stately waltz," a "majestic swirl," and a "primordial fireball"—images with wide traffic in such documentaries where they are frequently evoked to describe, for example, water bugs, herds of elephants, flocking cormorants, or perhaps a volcanic eruption, respectively. By means of "one master equation," Greene proclaims, the theorist-hero has rendered domestic all of these "riotous fulminations," an expression he repeats later on (135). This passage, as overwrought as it is, is not exceptional but exemplary. A heterogeneous imaginary that gestures toward romantic quest, wilderness safari, depth penetration, and tapestry weaving informs the entire account, influencing nearly every specific choice of image. The relative obscurity of the abstract theoretical space of string theory technical exposition has become embellished in the popularization with a complementary romantic imaginative regime. Greene uses the "poetic license" he grants himself through the aegis of pedagogy to displace technical heterogeneity into an imaginative pastiche, which is comparatively heterogeneous but seemingly less so due to the currency of its constituent imagery.

Echoing Kaku, Greene later asks a rhetorical question, "Why strings? Why not little frisbee disks? Or microscopic bloblike nuggets? Or a combination of all of these possibilities?" (20) Kaku substantiates the image of the string by declaring its "natural" affinity with the organic strings of protein and DNA that compose the human and animal body. Greene summons the canonical analogy of music—that finds its pedagogical lineage reaching back through Veneziano to Euler in his mathematical study of the violin string.[16]

> Music has long since provided the metaphors of choice for those puzzling over questions of cosmic concern. From the ancient Pythagorean "music of the spheres" to the "harmonies of nature" that have guided inquiry through the ages, we have collectively sought the song of nature . . . With the discovery of superstring theory, musical metaphors take on a startling reality, for the theory suggests that the microscopic landscape is suffused with tiny strings whose vibrational patterns orchestrate the evolution of the cosmos. (135)

The conceit that nature produces music is a classic anthropomorphism with a broad cultural currency that, as Greene notes, dates back to antiquity. It is anthropomorphism because music is entirely a cultural product. In the imaginary, nature exhibits human-like features in its production of music-like phenomena. While Greene acknowledges that this creative projection is metaphoric, he nevertheless playfully undermines any careful distinction between a cultural product and natural phenomena when he boldly asserts that through string theory, "musical metaphors take on a startling reality." Presumably this is because string theory is not something constructed per se, but a "discovery." Inasmuch as music is a cultural product with aesthetic and pragmatic aspects, so too is the "song of nature," since it too possesses "harmonies" that manifest an imagined purpose, in this case, the "evolution" of an orderly cosmos. It is precisely this playful blurring—made conscientiously tentative in the use of qualifiers such as "suggests" and "according to"—of the boundaries between culture and cosmos that gives strings as natural phenomena an imaginatively palpable reality.

Greene presents an image of the microcosm explicitly as a "landscape"; that landscape "is suffused with tiny strings." The image cleverly plays on the amphibology between object- and location-event schemas that also informs the imaginaries of string theory technical exposition. One may imagine the verb *suffuse* as either the spreading of a liquid, light, or color over a surface or as the permeation of a volume by a substance. The substance, in this case "tiny strings," thus suffusing the "landscape," can be either distinct from the landscape-as-surface or of the landscape-as-volume, both part of the location and/or an object that stands out from it in a relationship of figure to ground. When Greene introduces another imaginative frame—a journey to the "microscopic landscape"—the object-event schema is emphasized, and one encounters "tiny strings" as objects that are figures contrasted to a ground. Then, riffing on the parallel between superstrings and the strings of a musical instrument, Greene states that their "vibrational patterns orchestrate the evolution of the cosmos." Just as musical instruments generate notes, which together may comprise an orchestral symphony, so too do the "notes" of superstrings generate the cosmos-as-symphony. In this analogy, figure and ground are transposed to emphasize the location-event schema. The "microscopic landscape" now becomes foregrounded as "vibrational patterns" that make up matter—the form of the cosmos. Accordingly, in the reversal, strings are the substance beneath that material form, generating it. Of course, Greene does not expect the reader to imagine these shifting scenarios in exactly literal terms. If anything, the imaginary does its work all the more effectively if it is passed over quickly and simply taken, to echo Le Doeuff, as "the way it is." Even though the amphibology between string as underlying substance and/or string as object on a ground ruptures any implied coherence to Greene's imaginary, the sheer onslaught of images and the mobility of the authoritative voice as it jumps from one imaginative frame to another conspire to generate a finely blended feeling of both familiarity and wonderment, to which the reader, as a precondition of access, is expected to submit.

Shortly thereafter, Greene offers another definition of the string.

> According to string theory, the elementary ingredients of the universe are *not* point particles. Rather, they are tiny, one-dimensional filaments somewhat like infinitely thin rubber bands, vibrating to and fro. (136, emphasis in original)

In this passage, the first image presented is that of "elementary ingredients," framed within the epistemological conceit of appearance and essence. This image of strings as "elementary ingredients" of the universe bears with it traces of the ambivalent hybridization of the man-made and the natural, as in string theory technical exposition, where the string undergoes an imagined transformation from the quasi-mechanical—that which is constructed through mathematical procedure—to the autonomously natural, that which is discovered. In technical exposition, the intersection of a procedural space with an abstract theoretical space instantiates an encounter of theorist with strings as natural phenomena. There, theorists imaginatively isolate a select component embedded within mathematical argument and label it as something that is categorically basic. Accordingly, in technical exposition the image of the string comprises an "ingredient" in that its naturalness is motivated by human-scale qualifiers such as graspability and quotidian familiarity. Conversely, the image of the string is "elementary" because its mathematical consistency engenders its legitimacy as natural phenomenon. Meta-discursive epistemological assumptions on what constitutes a natural object are unacknowledged as such and thus uncontested. Here in a popularization, a pedagogic space that intersects with the abstract theoretical space of string theory provides an opportunity for a more overt insistence on those underlying assumptions. The accessibility of a popularization, in this sense, provides the theorist with a means to reify conceptual content—to more thoroughly figure it in conventionally concrete terms—within a pedagogic context where underlying epistemological assumptions are non-controversial since they come prequalified as part and parcel of the "appropriate simplification" at stake, and as such, merely ludic.

Greene further concretizes the string by evoking images of "tiny filaments" and "rubber bands," reminiscent of Susskind's "ideal rubber band" in "Dual-Symmetric Theory of Hadrons—I." Here again the logic of image selection, while affording access, betrays a certain amphibology in that access. In general, a filament may denote either the organic—a vegetable or, in some cases, animal structure—or something artificial, such as the tungsten filaments in incandescent light bulbs. In the case of the latter, one may imagine the string of string theory glowing with energy much as a light bulb.[17] The filament is like the notion of the field in the nineteenth-century electromagnetism of Faraday and Maxwell, an image the two theorists borrowed and abstracted away from a contemporaneous agricultural currency.[18] While categorically basic, the filament is tangible enough that it calls to mind a spectrum of images from concrete to abstract, from natural to agricultural to mechanical. Again, it is an image that carefully maintains the ambiguity between construct and nature. The image of the rubber band, on the other hand, is unambiguously artificial and, as such, Greene distances it from the imaginary through the qualified simile of "somewhat like."

A coherent image of a rubber band is further undermined by qualifying it as "infinitely thin." As I mentioned in the previous chapter, according to Lakoff and Johnson, to imagine the infinite engenders the extrapolation of an indefinite process: an object moving in a linear space-like time dimension terminates in a bounded place-as-destination which, importantly, is marked as unimaginable. The reader is expected to imagine "tiny" filaments as objects that lie at the boundary of sight (one must squint to see the tiny), and as such, it is a liminal image restlessly situated between an extended object and a point. Accordingly, the point extended in one dimension to become a line induces a sense of thinness that is also at one's perceptual limits; for example, the drawn out mark of a pencil on a page. But in this case, one imagines a prototypical thinness becoming even thinner indefinitely into an unimaginable end-place of infinite thinness. This extrapolated destination is unimaginable because, strictly speaking, for a rubber band to be infinitely thin it would have to have absolutely no thickness at all, and thus be nonexistent. When Greene evokes the infinite here the reader is meant to take it at face value, even though in other, human-scale contexts, Greene would undoubtedly aver, such an image would be "nonsensical." It is at this point in the guided imaginative exercise that the reader is simply meant to take the author's word for it—a submission to what Le Doeuff calls "straightforward dogmatization" (12). Substantiality predicated on human-scale imagery defers to access made tenable by means of Greene's authority as an accomplished string theorist and pedagogue.

So, again, one finds a certain contradiction—a crisis point where, as Le Doeuff puts it, there occurs a "rework[ing of] elements of a mode of discourse which [scientific discourse] elsewhere repudiates" (5). Distanced from its conceptual source, a popularization may exploit the inferential resonances of the epistemological imaginary through which it constitutes itself. In technical exposition, overt invocation of such epistemological conceits is obviated by its juxtaposition to its mathematical double—and furthermore, by the encompassing intertext of professional culture. In a popularization, a farrago of "down-to-earth analogies," at once human-scale and duly authorized by an authority, stands metonymically for the cosmos. The popularization is then wholly self-referential, like an uroboros. The text relentlessly folds the pedagogical space back into the abstract theoretical space it has exposed, all the while fully drawing on the stock images, conceits, and conventions of the genre. Greene assures the reader that while he has "tried to stay close to the science," he has also, contradictorily "avoided technical language and equations" (xi). In effect, this is an inadvertent admission to the effect that he has accomplished this feat only insomuch as he has hewed closely to an imaginary that, from a strictly realist perspective, in no way constitutes the conceptual content of technical discourse. I make this point not to corroborate Sokal and Bricmont's accusation that popularizations distort, but to suggest that the distortion is everything. The only means by which we may comprehend as coherent spaces and phenomena impossibly remote from human scales are by means of an imaginary. A popularization such as *The Elegant Universe* does not distort, but reorients, reintegrates, and repackages an imaginary in order to do more than simply disseminate the science of string theory. Science, in effect, becomes a cosmic order, and as such, implies a set of appropriate attitudes and behaviors to hold with respect to that cosmos. A social order follows directly from our imagined place within it.

Randall's *Warped Passages*

Randall begins the first chapter of *Warped Passages* thus: "The universe has its secrets. Extra dimensions of space might be one of them" (1). This notion that nature is secretive jibes readily with the epistemological conceit of appearance-as-surface and essence-as-depth—where a hidden essence must be found out—that had an easy traffic in both Kaku and Greene. To justify the image, Randall contends that "as physics evolved in the twentieth century, it moved away from things that can be directly observed with the naked eye to things that can be 'seen' only through measurements coupled with a theoretical train of logic" (9). In large part, this stock conceit of surface-as-appearance and depth-as-reality depends on the folk theory in Anglo-American culture that it is possible to "directly observe" the world with the "naked eye." In its ordinary sense, direct observation simply means to view an object or event without the aid of instruments. One gets the impression that Randall might be imagining the likes of Tycho Brahe, the sixteenth-century Danish astronomer famous for his accurate and comprehensive astronomical observations. Since the telescope had yet to be invented, conventional reminiscences of Brahe tend to imagine him gazing up at the night sky and meticulously noting what he "directly" observed.

Yet even Brahe constructed and made use of a number of mediating instruments, including astrolabes, quadrants, orreries, and armillaries, among others. In making his observations, he also consulted a substantial body of astronomical data from his various Greek, Arab, and European predecessors. This decided lack of nakedness with regard to the eye was compounded by the fact that the heavenly bodies had to be observed over lengths of time in order to account for the full scope of their cyclical trajectories. This diachronic lag required even ancient astronomers to employ a "theoretical train of logic."[19] The image of the naked eye, unaided by instruments, reinforces the conventional assumption, thoroughly discredited by contemporary cognitive science, that perception and cognition can be isolated from each other. This assumption is an aspect of what Lakoff and Johnson call "the Folk Theory of Faculty Psychology: a model of the mind as divided into discrete 'faculties'" (409).[20] In fact, what the "naked eye" sees is always mediated by instruments, whether external to the body, in the form of astrolabes, telescopes, spectrometers, or colliders, or internal to the body, in the form of integrally looped and mutually constituting perceptive-cognitive neural processes. Words that stand for things that are categorically basic, such as *apple* or *star*, are understood as available to the naked eye because they are long-standing, nearly universal, and fully conventional abstractions. The architectonic primitiveness of such basic images allows a culture to substantiate them as purely objective, as real things in the world with inherent rather than interactional properties.[21]

Theorists envision themselves intervening in a cosmos that is imagined as distinct from them, yet our bodies also inhabit the cosmos as an integral part of it. With respect to string theory, interaction between physicist and cosmos occurs through the orienting mediation of mathematics (and in ancillary, through instruments). The mathematics of string theory serves as a form of proprioception—an abstracted propositional system of differential relations and highly precise prompts for movement that orient embodied interventions into and with the world. In string theory technical discourse, mathematics is coupled to an imaginary, and concomitantly, an abstract theoretical space is coupled to a system (or systems) of mathematical proprioception. This coupling of mathematics with an imaginary generates a collective hallucination—a given culture's provisionally unified and taken-as-coherent cosmic order—by means of which it may regiment a relatively consistent intervention into the cosmos. The coherence of this cosmic order lies wholly in the acceptance of a particular cache of conventionalized imaginative conceits: appearance-as-surface, reality-as-depth, theoretical procedure-as-journey (where construct gives way to discovery), and basic cognitive categories (such as strings) granted the status of objective natural phenomena. An imaginary may be apt in that it affords an experience of relative coherence and substance to a mathematical regime, but it is never real in the sense of in "direct" contact with the world.

Throughout *Warped Passages* Randall is not concerned with strings per se, but rather with extra dimensions and the objects out of which they are composed—branes. From the outset, Randall attempts to clarify what she and other theorists mean by the concept of the extra dimension and its relationship to space. In chapter 1, "Entryway Passages: Demystifying Dimensions," she offers the following definition: "The number of dimensions is the number of quantities you need to know to completely pin down a point in a space" (13). She then goes on to define "space" as "the region in which matter exists and physical processes take place. A *space of a particular dimension* is a space requiring a particular number of quantities to specify a point" (15, emphasis in original).[22] Randall repeatedly reminds the reader that the world in which the body moves has three dimensions. It may move back and forth, left and right, and up and down—the dimensions of length, breadth, and height. As a result, the imagination, originating as it does in embodied experiences in these three dimensions, can only render objects and events in three spatial dimensions (or less).

In attempting to imagine extra dimensions, then, one is confronted with an impasse. Randall points this out when she writes that "the phrase 'extra dimensions' is especially baffling because even when we apply those words to space, that space is beyond our sensory experience" (12). Her solution to this conundrum is to suggest that a "multidimensional space might be an abstract one, such as the space of features you are looking for in a house, or it might be concrete, like the real physical space we will soon consider" (13). Randall continues: "when buying a house, you can think of the number of dimensions as the number of quantities you would record in each entry in a database—the number of quantities you find worth investigating" (13). Such a multidimensional space is abstract in that it contains more information than one would need to "pin down" a location in the three-dimensional space of our world.[23] It is telling that Randall chooses the image of the house to fortify her argument. As Bachelard claims in *The Poetics of Space*, "The house . . . is a privileged entity for a phenomenological study of the intimate values of inside space . . . for the house furnishes us dispersed images and a body of images at the same time" (3). As a matter of basic cultural literacy, we are all intimately familiar with the space of the house. The body moves with ease in the three dimensions of a prototypical house, while the imagination also moves with ease within the multidimensional abstract space (by Randall's reckoning) of all the feelings one experiences within the house as one occupies it. In a prospective new house,

Randall designates such an abstract space as the "quantities you find worth investigating"—a complex of physical features and associated affect-laden evaluations of those features. As a consequence, such an abstract space—the space of one's imagined and felt relationship to the physical space of the house—takes on a certain cohesion.

Pedagogically, Randall's association of a multidimensional abstract theoretical space with the comforting image of the house-as-space seeks to override the former's counterintuitiveness. By conceding to the multidimensional "features" of the house-become-home, the second part of Randall's solution may slip past the reader's attention.[24] If a multidimensional space is "beyond our sensory experience," then how can it be "concrete" in the ordinary sense of that word? The body may only come into contact with that which is accessible to its "sensory experience." If, by her own definition, the body inhabits only the three spatial dimensions of length, breadth, and height, how may one know the multidimensional space that she alludes to as "real physical space"? Would that not also be an abstract space, comparable to the abstract theoretical spaces of, for instance, the heterotic superstring? Extra dimensions may be concrete only in an abstract sense: inasmuch as they are made up of particles (such as gravitons) and thus have mass can they be understood as being concrete. The status of concretion functions wholly within the context of a theoretical definition, rather than through its substantiation by direct contact with the body. Her proposed solution to this paradox relies on the epistemological conceit of the naked eye. By projecting an imagined seeing (an eye freed from the body) into the abstract theoretical space, Randall establishes the epistemological preconditions through which that space may become categorically real. To argue that experiments may eventually confirm the "existence" of extra dimensions is only then conveniently to ignore the crucial mediating of the body between the observer—imagined as imagining—and the probing instrument. The "nakedness" of the eye here is in its idealized purity of observation, freed from the constraints of the body. Were the human body to intervene in such a multidimensional space, that space would be known in terms of the body, i.e., three-dimensionally. Figuring extra dimensions in a homelike context effectively whitewashes this epistemological paradox.

In chapter 15, "Supporting Passages: Brane Development," an epistemology that relies on an idealized "naked eye" becomes more overt. Recall that strings are generically defined as one-dimensional extended vibrating objects; accordingly, branes are two- or more-dimensional objects that also vibrate. Randall describes the brane thus:

> We will now see that branes are more than just a location; they are objects in their own right. Branes are like membranes, and, like membranes, they are real things. Branes can be slack, in which case they can wiggle and move, or they can be taut, in which case they will probably sit still. And branes can carry charges and interact via forces. Furthermore, branes influence how strings and other objects behave. (305–6)

In one imaginative frame, a brane constitutes the entirety of the abstract theoretical space—a location-event schema. But in the passage, Randall reverses figure and ground within the imaginary. The brane-as-space, taken as a contiguous whole, becomes an object contained within a higher dimensional background space—the bulk. As such, "branes are more than just a location; they are objects in their own right." Employing the same reversal of logic as in the case with extra dimensions, Randall then asserts that "branes are like membranes, and, like membranes, they are real things." Within the imaginary, the *and* in this statement would seem to stand in for a *because*—a simile becomes causative. In effect, Randall seems to be arguing that branes are real *because* they are like membranes. The imagined substantiality of the brane—its reality—depends on a sufficient familiarity with other human-scale membranes. That Randall speaks with the full resonance of professional authority ultimately resolves this quandary of ambivalent coherence: branes are real simply because she claims they are so.

The description that follows of the brane's attributes exploits an implied intimacy with membranes in human-scale experience. A juxtaposition occurs in the next sentence that conflates an impression of branes as being both drum-like *and* animal-like: they can be "slack" or "taut," while they can also "wiggle" and "sit still." The succession of imagined attributes establishes a tension between the constructed and the animal-like or zoomorphic. The adjectives *slack* and *taut* imaginatively evoke the image of a drum, something manmade; it is crafted from parts and, as a whole, articulated to perform a function. Like the string and the violin string, there is a strong association within the physics pedagogic tradition between membrane and drumskin, dating back to Euler's work on the subject, which he conducted as a follow-up to his work on violin strings (Stewart 66–67). On the other hand, it is the choice of the verb *wiggle* that implies the brane's animal-like behavior drawing, as it does, on the subtle association between wiggling as a particular autonomous motion and living things such as snakes, worms, and insects. While inanimate objects may very well "sit still," its juxtaposition with "wiggle" serves to emphasize the brane's autonomous behavior.

Next, Randall states that branes can "carry charges" and "interact via forces." The notion of carrying a charge is tautological in that charge literally means "to carry a load." In the mathematics of physics, charge signifies a number (a scalar) or set of numbers (a vector or tensor) at a particular point in a multidimensional abstract theoretical space. In this context, the image of an object carrying something, coupled to an image of interaction via "forces," contributes to a sense of a deterministically mechanical phenomenon. But then, once again, the machinic surrenders, albeit abstrusely, to the zoomorphic in that branes "influence" the "behavior" of "strings and other objects." To write of "influencing behavior" with respect to these theoretical phenomena hints at a certain unpredictability that exceeds the conventional determinism of the strictly mechanical. In Randall's imaginary, branes exceed construction, behave autonomously, and thus are real. Access to branes then depends on the reader's willingness to accept their characterization in multivalent terms, terms that play off of a wide range of associations.

Yet Randall retreats from a more blunt zoomorphic characterization when she reemphasizes the corollary of brane and drumskin.

> Branes in string theory have finite tension. Brane tension is akin to the tension of the surface of a drum that returns to its taut position after you pinch it or punch it. . . . Because the tension of branes is finite, branes can move and fluctuate and respond to forces, just like any other charged object. (307)

Just as one may physically "pinch" or "punch" a drum, the reader may imagine doing the same to a brane. This imagined interaction, Randall argues, is possible due to the brane's surface "tension." Like a drum, "branes can move and fluctuate and respond to force." But then Randall further explains that "the feature of branes that is essential to their potentially observable applications is that they can trap particles and forces" (322). Now with this "essential" feature, the analogy between brane and drumskin loses its aptness. Just as she and Sundrum did in "An Alternative to Compactification," Randall employs an image of a pellicular, semi-permeable container. The brane "traps" most matter and force particles but allows certain other force particles—specifically, gravitons—to pass through its boundary into the multidimensional metaverse, or bulk. Randall continues:

> Certain branes always have particles and forces that are trapped on them. Like housebound cats that never venture beyond the walls of

> their domicile, those particles that are confined to branes never venture off them. They can't. Their existence is predicated on the presence of the branes. (323)

Interestingly, Randall evokes the home once again in her simile where particles are equated to "housebound cats" and the brane is a "domicile" whose "walls" stand for the brane's boundary. Particles are "confined" and "never venture off." The use of the adverb *off* is noteworthy here in that it hovers between the sense of the brane as two-dimensional drumskin *and* as three-dimensional house-as-volume, since a cat, strictly speaking, would not venture *off* of a house, but rather, *out* of it.

This amphibology bears a comparable tension to the image of "trapped gravity" as object- and/or location-event schema in "An Alternative to Compactification." In the technical exposition on which this popularization's description is based, the imaginary shifts between two contradictory frames. In one, particles are integrally part of the brane—as vibratory attributes of its very substance—a location-event schema. In the other, particles are objects distinct from the brane, and the brane, like the surface of a container, traps them within its boundaries. With the example of the housebound cat, Randall deploys a double analogic mediation. Within technical discourse, the expository imaginary stands in analogic relation to mathematical argument. In this passage, the simile of housebound cat stands in analogic relation to the technical imaginary—brane as object and/or location. What in technical discourse is an encounter with natural phenomena in a remote, alien space gains further imaginative traction through an even more accessible domestication—in this case, literally, by likening the particle-brane relationship to that of a cat with a house. The imaginary then shifts perspective in that the particle's very "existence is predicated on the presence of the branes"—a notion certainly not appropriate to images of house cats or the punching of drumskins.

Like technical exposition, the imaginary relies on a macaronic medley of imaginative frames and schemas. Yet unlike technical exposition, the privileging of accessibility within *Warped Passages* engenders a more insistent return to the homelike. But this further degree of domestication is not necessarily a case of conspicuous simplification. Rather, as with Kaku and Greene, it constitutes a displacement of complexity away from the abstract theoretical space of technical discourse into a more fully articulated mythopoeic space that serves to justify and substantiate in an expanded imaginative register the remote and monstrous phenomena that are branes and extra dimensions. In the popularizations considered, monsters become fully de-monstrated. Through the intersection of procedural-become-pedagogic space with an abstract theoretical space, the pedagogic reimagines the technical while also exceeding it in a way that cannot be dismissed as mere simplification or distortion. In spatial terms, the access of popularizations is a coming back that imaginatively mirrors the going out of the encounter of technical discourse. But access adds its own complexity to the imaginary—a proliferation of quasi-domestic imagery within the heretofore mysterious, hidden things of the cosmos. This further domestication, in turn, provides supplemental scaffolding for the elaboration of string theory as a cosmic order.

Accessibility and Authority

Adapting the ideas of Stephen Hilgartner and Thomas Gieryn to physics popularizations, Felicity Mellor argues:

> The rhetoric of accessibility implies twofold boundary work. On the one hand, it assumes that a boundary exists between science and non-science which isolates science from non-scientific publics and which determines "accessibility" as a problem to be addressed. On the other hand, it claims that these popularizations, by virtue of their accessibility, are able to dismantle or overcome the boundary between science and non-science. (516)

What Mellor highlights here is how the "rhetoric of accessibility," by imaginatively framing science as a bounded space, attempts to justify popularizations precisely because they transform the boundary between science and non-science into a third space. This transformation is akin to a Venn diagram where technical and popular discourses, as epistemological domains imagined in spatial terms, overlap to form a distinct space that is scientifically authoritative yet also accessible. In the necessarily *uni*directional dissemination of scientific knowledge from technical to fully popular discourses, the boundary between the two now becomes a demarcated zone—what I have called a *pedagogic space*. In keeping with Mellor, although the selected popularizations promise to provide access to the abstract theoretical space of string theory, they also serve to reinforce the boundary between such a space and, by implication, a more broadly delineated everyday social space, since they omit key content from the technical discourse and implement a set of implicit provisions for access to what remains. First of all, it must be a guided and, as a consequence, a restricted access.

As discussed earlier, Hilgartner argues that the flexibility of this boundary between science and non-science depends largely on whether writers of popularizations are able clearly to distinguish between what he calls "appropriate simplification" and "distortion." In turn, Mellor suggests that the claim to appropriate simplification protects "the position of science in a hierarchy of ways of knowing while appearing to be merely playing the popular market" (519). Because popularizations constitute, as she puts it, "a relatively uncontroversial context," this, in certain respects, means that "covert" boundary work can "go largely unchallenged" (519). Popularizations are uncontroversial in that they do not comprise the actual science itself. Rather, they are a self-avowed representation of science for educational and entertainment purposes, where the authors themselves (more often than not leading scientists in a given field) determine what is appropriate simplification in accordance with the generally accepted conventions of the genre. Mellor argues that in this cultural climate, popularizations become a form of "routine" boundary work. She contends that this "routine boundary work is significant because it maintains a cultural resource of normative images and understandings of science which acquire a wide public circulation and can be invoked whenever challenges are made to the position of science in society" (521).

While string theory popularizations reevoke the tension between construct and natural phenomenon that one finds in the imaginaries of technical exposition, in a wider cultural context, the license to play that popularizers grant themselves serves to distract scrutiny away from the imaginative content of technical discourse. All imagery becomes marked as simplification, whether it is appropriate, as is the simplification that falls within the purview of pedagogy, or inappropriate, that which is deemed distortion. It is easier to judge a popularization as a distortion when the "whimsical" imaginary therein is held up to the standard of the "hard" conceptual content of technical discourse, to that is, mathematical argument and its accompanying precisely defined technical jargon. With all "simplification" shunted off to popularizations, the expository content of technical discourse may go unchallenged in its claims of conceptual purity. It benefits from a comparison to popularization that is bound to be, as a matter of epistemological policy, favorable. Access obscures and suppresses the imaginative contingencies of technical exposition, thereby legitimizing the cosmic order presented within technical discourse all the more as something purely objective and natural. In the economy of boundary work, the more objectivist hard-liners trivialize popularization, the more they relieve technical discourse from the need to examine an epistemological imaginary that substantiates it.

Mellor is principally concerned with the role that popularizations play, as a particular form of routine boundary work, in a politics of self-preservation. Popularizations effect the further institutionalization of a given scientific practice through the wide "circulation" of "normative images" in the culture that supports it. The popularizations considered in this chapter may very well protect the position of string theorists within society, as Mellor suggests. My concern here has not been with sociological causes, but rather with the specific ways in which string theory popularizations generate a novel imaginary and then condition it for that wider cultural circulation. This involves a twofold process. Firstly, the theorists enmesh the abstract theoretical space of strings, branes, and extra dimensions with a supplemental pedagogic space predicated on various stock epistemological conceits. Most notably, they are: string theory as essence as opposed to appearance, an allotropy where many forms reduce to one substance, a depth below a surface, a correct reading or decipherment of the book of nature (i.e., nature as textual code or information), the goal of a quest or the destination of a journey outward, a homecoming (however circuitous), and the exploration and eventual domestication of a wilderness-like space. These epistemological conceits are not exclusive to string theory, but rather emerge from a long-standing cultural tradition that the theorists simply take as given. The very conventionality of these epistemological conceits lends legitimacy to the more specifically string theoretical imaginary that they encompass.

This paves the way for the second part of the process of access. In technical discourse, basic images, duly encountered and quasi-domesticated, become "discovered" autonomous natural phenomena. Through popularizations, string theory imagery becomes more readily accessible, thus mobile and consumable. This is because the radical, and in many respects, utterly incoherent heterogeneity of the string theory technical imaginary becomes tamed by analogy to ever more human-scale and domestic contexts. An imaginary in a popularization is no more or less heterogeneous than a technical imaginary—and thus no more or less incoherent. But that heterogeneity is displaced away from technical complexity into the playfulness of pedagogic monologue. The relative incoherence that results is acceptable for two reasons. Firstly, because of its very conventionality, its appropriate blend of the familiar and the weird, the known and the new, and secondly, due to the authority of the theorist that vouches for it, which serves as an adequate substitute for the absent mathematical argument. The imaginary of string theory popularizations finds its own genre-specific and conventional balance between, to revisit Serres, a rapport with things and relations among us. As such, in the transmutation of a string theory imaginary from encounter to access, it becomes the basis for a reinforced resonance between a particular cosmic order and social space.

As mentioned at the beginning of this chapter, accessibility is a notion that the three popularizers themselves embrace. Much like Greene in *The Elegant Universe*, in the preface to *Hyperspace*, Kaku claims that "this book makes available, for the first time, a scientifically authoritative but accessible account of the current fascinating research on hyperspace" (viii). This assertion creates the expectation that readers enjoy a privileged status in that they are the first to be privy to something hitherto esoteric and exclusive. But the more significant claim that Kaku makes for his text is that it is both "scientifically authoritative" *and* "accessible." While he does not explicitly reassure his readers that the text will omit the mathematics on which string theory technical discourse is based, the choice of the qualifier "accessible" suggests this. *Access* becomes a code word for omission of the properly technical. By offering access, Kaku reinforces the sense that the knowledge he is about to impart is bounded—and that those who would enter into this space need some special dispensation or guide. The dispensation comes from Kaku's assertion that what readers are about to consume, while being accessible, is also, nevertheless, "scientifically authoritative"—a seeming contradiction. Since readers have no access, both within the text and, presumably, due to lack of professional training, to the extratextual mathematics that constitutes the conceptual content of string theory, Kaku would have us accept the text's scientific authority on the basis of his pronouncement alone (which he later backs with credentials).

Hyperspace as a text mimics, in this respect, a faith-based transaction. The presumption is that a submission to authority will by itself adequately substantiate epistemological legitimacy. As such, this is not a form of empiricism but rather, pedagogy. Kaku explains:

> Because the hyperspace theory takes us far beyond normal, common-sense conceptions of space and time, I have scattered throughout the text a few purely hypothetical stories. I was inspired to utilize this pedagogical technique by the story of the Nobel Prize winner Isodore I. Rabi addressing an audience of physicists. (xi–xii)

Here Kaku specifically qualifies his notion of making string theory accessible in terms of pedagogy. In the logic of the pedagogic space of the text, he locates the reader in a place where he or she possesses "normal, common-sense conceptions of space and time," and then warns him or her that "the hyperspace theory" will take them "far beyond" these conceptions. With Kaku as guide, readers will be taken into an exotic realm that defies a "normal, common-sense" relationship with the world. He then evokes the image of scattering to describe his use of a "few purely hypothetical stories" that will serve as "pedagogical techniques." These techniques presumably are on offer to help orient the student-reader within the abstract theoretical space Kaku will expose and also to help "make clear" some of string theory's more counterintuitive concepts.

Through the use of an imaginative conceit of guidance to a place "far beyond," Kaku is placing his text within a cultural tradition that comes prestructured with images of space-as-place and its proper transversal. Also, those "purely hypothetical stories" function as pedagogical techniques precisely because they reformulate the alien space of string theory in terms of "common-sense" images. In this context, "common-sense" signifies one or more received imaginative conceits that have such a currency within the culture that they are simply taken as normatively real. Common-sense then has more to do with what is common—a kind of traffic, a cultural currency, of images—than what is (directly) sensed. What is abnormal and/or counterintuitive in "the hyperspace theory," then, lies in novel recombinations that contest a conventional "common-sense" imaginary, not in the constituent images themselves, nor the technical knowledge for which these images double, nor some universal unmediated sensory experience that transcends culture.

Early on in the preface to *The Elegant Universe*, Greene also establishes the parameters of access:

> I wrote *The Elegant Universe* in an attempt to make the remarkable insights emerging from the forefront of physics research *accessible* to a broad spectrum of readers, especially those with no training in mathematics or physics. Through public lectures on superstring theory I have given over the past few years, I have witnessed a widespread yearning to understand what current research says about the fundamental laws of the universe, how these laws require a monumental restructuring of our conception of the cosmos, and what challenges lie ahead in the ongoing quest for the ultimate theory. (x, emphasis my own)

This passage compactly expresses Greene's position with respect to access, authority, and the epistemological. In the first sentence Greene conflates two conceits completely in keeping with those of Kaku: the image of scientific knowledge as "insight" into an abstract theoretical space where the surface represents "common-sense" appearance and the interior or depths are marked as reality. This movement inwards/downwards is compounded by the impression of movement outwards to a "forefront," a liminal zone between the domestic and, inferentially, a wilderness-like space. Like Kaku, Greene offers to make this difficult material, for which direct engagement demands "training in mathematics and physics," "accessible." The notion of access capitalizes once again on the inferential structure of the complementary conceits of "insight" and "forefront," which each imply a semipermeable boundary that can only be crossed with the appropriate knowledge or guidance. Greene further coaches the reader to make the most of the proffered access to esoteric knowledge with the following advice: "the reader may need to pause now and then, to mull over a section here or ponder an explanation there, in order to follow the progression of ideas fully" (xi). He also, in certain "more abstract" sections, has "taken care to forewarn the reader about these sections and to structure the text so that they can be skimmed or skipped with minimal impact on the book's logical flow" (xi). Thus prompts for the reading of the text include pausing, mulling, pondering, skimming, and skipping. Greene is concerned with conditioning the reader to receive the text properly, in order to maximize its potential to "provide a startling and revolutionary perspective" (xi). The preparation for reception and revelation mimics, in certain respects, the ritualized structure of a religious conversion, which also calls for a certain ordering of practices that are to be carefully executed so that the new, transcendent perspective may be properly experienced.

Greene takes pains to establish his credentials through the casual mention of "public lectures" he has given—a role generally reserved for the statesmen of physics. As with Kaku, only later in the exposition does he present the accomplishments that garnered for him a professional reputation.[25] Greene declares that he has "witnessed a widespread yearning to understand" amongst those who have attended these popular lectures. This image is almost evangelical in its rhetorical inflection; his is a sacred duty to bear witness to those yearning to come out of ignorance and, ultimately, to satisfy that yearning. Accordingly, Greene declares that his aim is to "both enrich and satisfy this curiosity" (x). In effect, he expects, however tacitly, that the reader's response to his exposition will first be one of yearning and then curiosity slaked, of gratitude for the enrichment offered.

As with Kaku and Greene, Randall frames her work as a theoretical physicist in terms of an epistemological "search" for that which lies underneath the surface of appearance. Like Kaku and Greene, for Randall, model building is "adventure travel through concepts and ideas" (8). Moreover, she structures her "faith" in extra dimensions as real phenomena in the form of an epiphany.

> Do I believe in extra dimensions? I confess I do. . . . One day on my way to work about five years ago, as I was crossing the Charles River into Cambridge, I suddenly realized that I really believed that some form of extra dimensions must exist. I looked around and contemplated the many dimensions I couldn't see. . . . Greater familiarity with extra dimensions has only increased my confidence in their existence. (3)

A mythopoeic religious imaginary frames this anecdote. Firstly, it is a "confession," borrowing the imaginative structure of the juridical ritual of Catholic orthodoxy. Her "realization" occurs on a bridge as she crosses over in her commute from home to work. This image of a crossing over suggests the transformation that occurs in a conversion experience, where the initiate makes the passage from a space of unbelief to a particular devotional camp, marked in the scene as two opposing riverbanks. The epiphanic flash then transmutes into a meditative "contemplation" where the devotee now has the power to see immanently that which was heretofore hidden from view. A genuine encounter with the mystery of extra dimensions demands submission to its insistent reality. And then later, "greater familiarity" steels her faith in their existence. This sense of familiarity speaks directly to the notion of quasi-domestication. Active imaginative engagement with extra dimensions through her work has led inexorably to just such a feeling of comfortable familiarity, steeped, as it necessarily would be, in an imaginary populated with abstracted recombinations of everyday objects and events, of a homelike space transmuted and extended to a procedural space made accessible. In the logic of the imaginary, Randall has brought her home with her across the river. And Randall's confession of conversion affords us access to a comparable (yet subordinate) conversion experience.

Borrowing a methodology from Eric White, in *Reading Popular Physics*, Elizabeth Leane analyzes Stephen Hawking's *A Brief History of Time* and Steven Weinberg's *Dreams of a Final Theory* in terms of their respective "emplotments"—where the narrative arc of these texts may be classified as comedy, tragedy, romance, or farce. According to the classification that White and Leane draw on, comedy and tragedy are "closed" forms that end with the resolution of a conflict. Romance and farce are open forms, constituted by an endless sequence of events. Whereas in romances a purpose or goal drives plot, farce is typified by a "chain of accidents." Leane argues that *A Brief History of Time* is comic; *Dreams of a Final Theory*, farcical (127–31). With respect to string theory popularizations, that an emplotment structures a given imaginary helps to explain how cosmic order and social space become closely coupled. As the previous chapter showed, the open-ended, purpose-driven structure of the technical imaginary most closely resembles romance. The imaginative access afforded by the three popularizations this chapter has considered is also open-ended and goal-oriented. All three see purpose in string theory and model building, and all three see the goals endemic to that purpose unfulfilled yet worthy of a continued pursuit.

In *Anatomy of Criticism*, Northrop Frye defines the hero of romance as one who is "superior in *degree* to other men and to his environment" and "whose actions are marvelous but who is himself identified as a human being" (33, emphasis in original). The three authors portray themselves and other historically significant contributors to the development of string theory in realistically human terms; there are no superhuman abilities or exploits on display. Yet an economy of status elevates these contributors to varying degrees of heroic superiority. In their hieratic pantheons, certain theorist-heroes, especially in Greene's account, warrant the status, due to their "marvelous actions," of genius—in particular, Newton, Einstein, and Witten. Furthermore, in the world of these string theory popularizations, the "ordinary laws of nature" are indeed "slightly suspended," as Frye has it, insomuch as the established theories of the quantum and general relativity—as stable, institutionalized expressions of scientific knowledge—are upended to make way for a "revolutionary" new theory, string theory (33). Accordingly, at the "forefront" of both the cosmos and scientific knowledge itself as abstract spaces, the theorist-heroes of string theory possess the capacity to suspend those conventional laws through their insights.

Within the imaginary—whose "postulates of romance," as Frye puts it, are firmly established at the outset of the exposition—the theorist-hero makes use of "enchanted weapons" in the form of an esoteric mathematics that is evoked but not deployed, such as differential calculus, Lie algebra, perturbation theory, or noncommutative geometry (33). The situating of the specter of mathematics on the margins of these texts invests it with a certain "talismanic power" (33). Indecipherable to the non-specialist lay audience, mathematics, fleetingly referenced, presents "figures or characters, to which are attributed occult powers" (*OED*). This is not to suggest that a lay audience literally views mathematics as magic, only that a marginalized mathematics mimics the magical insomuch as it exhibits a functional yet occluded power within the imaginary of the popularization. That power is occult in that it is indecipherable yet nevertheless taken to be effective—an arcane language that properly invoked by the initiated (the author and his professional colleagues) has the power to divine an essential reality previously hidden from view. Like a traditional talisman, this power is cosmic in its scope. The talismanic immanence of mathematical argument, banished by the author at the presumed behest of a mathematically challenged lay audience, reinforces the authority of the author in his priestlike ability to move with fluency across the boundary into the third space between technical and popular discourses.[26] Technical discourse thus becomes a supplementary, yet marginal "textual presence" that brings into sharp relief, for the lay audience, the abstract theoretical space of string theory as mysterious, wondrous.

An imaginary that couples a cosmic order to a particular social space gains its currency through promulgation by an authority. In string theory popularizations, that authority is the theorist-hero. As custodians of cosmic mysteries, Kaku, Greene, and Randall feel obliged to offer their readers an interpretation of the cultural significance of string theory as an abstract theoretical space in terms of relations "among us." They want to assist in helping us understand what string theory ought to mean to readers as non-specialists. The authors do this in large part by contextualizing string theory through the incorporation of denotative representations of what they deem important theorist-heroes and their pertinent exploits within the pedagogic space of the popularization. These denotative representations come to complement prescriptive prompts given by the authors to readers on how they ought to consume the text, and the imaginary therein. The conflation of a conventional epistemological imaginary with the abstract theoretical space of string theory evinces a "common-sense" that is both new and old. On the one hand, the authors set string theory, through a recombinant heterogeneous imaginary, in opposition to what they denigrate as an antiquated common-sense. On the other hand, the stock epistemological conceits with which they contextualize string theory serve to realign it back with the extant common-sense. The question then becomes: will this new, specifically string-theoretical common-sense, as it enjoys more and more circulation, condition those who participate in its dissemination and consumption to embrace the cosmic order implicit within its ostensibly whimsical pedagogic presentations?

The cosmic order latent within Kaku's imaginary strikes me as disappointingly puerile. He devotes the better part of the end of *Hyperspace* to speculations on what he repeatedly calls "mastery." In chapter 8, "Signals from the Tenth Dimension," Kaku writes:

> Assuming that some bright physicist solves the field theory of strings and derives the known properties of our universe, there is still the practical problem of when we might be able to harness the power of the hyperspace theory. There are two possibilities:
>
> 1. Wait until our civilization attains the ability to master energies trillions of times larger than anything we can produce today
> 2. Encounter extraterrestrial civilizations that have mastered the art of manipulating hyperspace (189)

Speaking for "our civilization" as if it were one cohesive whole, solving the "field theory of strings" would allow "us" to "harness the power of hyperspace," and thereby "master energies trillions of times larger than anything we can produce today." This is yet another image of domestication—where the notion of "power" draws on its potential to be "harnessed"—a contemporary abstraction of a primitive agricultural image. Later Kaku asserts that, "anyone, or any civilization, that truly masters the energy found at the Planck length will become the master of all fundamental forces" (269). Not one for understatement, shortly thereafter, he proclaims that "we have amassed more knowledge since World War II than all the knowledge amassed in our 2-million-year evolution on this planet" (274). These speculations are framed in terms of a peculiar version of romance. The purpose of string theory is understood as a chain of events: it begins with the solving of a monumental problem, then becomes an important contribution to the amassing of scientific knowledge, which, in turn, leads to mastery of the "tenth dimension" and, ultimately, the harnessing of unimaginable "power" for "our civilization."

Like Kaku, in *The Elegant Universe*, Greene appears to be preoccupied with fantasies of mastery and power. And like Kaku, the sophistication of Greene's imaginary would seem to reach a particularly acute limit when it attempts the integration of theoretical knowledge with a broader conception of a culture's relationship to the cosmos. Greene posits string theory, promisingly, as "a theory capable of describing nature's forces within a single, all-encompassing, coherent framework" (ix). In his vision, string theory would reduce our understanding of the universe to "one grand physical principle, one master equation" (5). Accompanying this dream of mastery is what seems to be a quasi-religious pietism of wonder and awe—directed both at the cosmos and at humanity's own power over it.

> It is truly inspiring that beings confined to one planet orbiting a run-of-the-mill star in the far edges of a fairly ordinary galaxy have been able,

> through thought and experiment, to ascertain and comprehend some of the most mysterious characteristics of the physical universe. (117)[27]

According to Greene, string theory provides the means to bring the remotely mysterious back within the fold of the "ordinary," the "run-of-the-mill"—thereby reinvigorating it. The ultimate human value appears to lie in the transcendence of our extreme existential marginality through the divination of nature's secrets. This image embodies the emotional extreme of a cosmic order of the human as isolated rational agent confined to an ordinary state of alienation within an opaque universe where liberation comes about solely through the demystifying clarity of an ever more purified rationality.

In the concluding chapter of *The Elegant Universe*, "Prospects," Greene writes of "grander surprises in store for us," of "wonder" and "marvel." He continues: "Whether any of our descendants will ever take in the view from the summit and gaze out on the vast and elegant universe with a perspective of infinite clarity, we cannot predict" (387). In Greene's imaginary, mastery of the cosmos through the access of "thought and experiment" yields an Olympian perspective awash in feelings of awe where nature has been conclusively found out, parsed out, contained. On one epistemological level, the access Greene offers is exhaustively panoptic, but on another level closer to home, it is not all that circumspect. String theory, as an imaginary, promises an explosion in the richness of information that space contains—the cosmos becomes a vibrating hyperspace of multiply entwined dimensions pulsating with energy. In the mastery of this richness, Kaku and Greene see, first and foremost, power. While the recombinant imaginaries of these two string theory popularizations may be novel, such a desire for power-mastery seems all too familiar.

To Randall's credit, it is more difficult to decipher any bluntly puerile desires lying in wait within *Warped Passages* than is the case with Kaku and Greene. As a self-proclaimed agnostic when it comes to string theory, perhaps she is less concerned with evangelizing string theory as a master equation bristling with futurist power. Randall's representations of social practice emphasize what she calls "connections," along with "personal relations," over images of the isolated theorist-hero on an epic quest with quasi-religious overtones, as Kaku and Greene would have it.[28] Yet in surveying the images of social practice within *Warped Passages*, a loosely organized pattern emerges—one that is best typified as a certain social conservatism, as a concern for an ordered society where all submit to the rule of law and where, concomitantly, as an outcome of this act of capitulation, its members become mutually comprehensible. In effect, a particular pattern of relations amongst "ourselves" serves to elucidate "our" rapport with branes—as things that stand metonymically for the cosmos.

As discussed in the previous chapter, Traweek describes the extent to which romance permeates the imaginary of informal high energy physics institutional discourses. Given its pervasive currency within this more cloistered domain, it is no wonder that romance would spill over into a popular discourse whose declared aim is to promote access for non-professional audiences. In its adaptation, the radical heterogeneity of the technical imaginary becomes displaced into a more socially resonant yet comparably heterogeneous imaginary. Such access—by its multiple acts of omission, profusion, and association—provisionally makes the boundary between science and non-science more porous. Paradoxically, though, access also tends to further retrench that boundary—in ways that the authors do not anticipate. String theory popularizations risk alienating non-theorists not by perplexing them with the weirdness of strings and branes, but rather, through an all too familiar reduction of physics into a cosmic order that champions mastery. If one accepts that a string theory popularization does indeed enact an emplotment that is open ended, but nevertheless feels compelled to reject the declared goal(s) of that plot, then string theory, as it is presented, gradually comes to feel rather quixotic. There may well be a risk that the readers of a string theory popularization became blinded by its dazzling images to the extent that they are ill-equipped to discern the more farcical aspects of these texts; in particular, their insistent return to stock epistemological conceits to substantiate an imaginary that is simply made to be real by authoritative fiat. In some respects, then, these string theory popularizations do not do full justice to the imaginative achievements of its technical discourse. Conversely, popularizations of this variety may very well expose the suspect romantic pretenses of the string theorists themselves.

CHAPTER 5

The Cosmic and Domestic

Adaptations of String Theory in Literature

What's more basic than string?
Cut string, you still have a string.
Like the Brahmin's idea of unity,
pouring milk into milk, which gives you
milk.
    Really, that's the opposite.
It's adding, not taking away.

—Elaine Terranova, "String Theory"

Functions of Borrowing

The previous chapter investigated the ways in which a selection of string theory popularizations makes an abstract theoretical space accessible through a pedagogical space. Within these string theory popularizations access becomes an imaginative mode that works to distinguish what theorists mark as novel in that space; in particular, images of strings and branes, from what the theorists frame as established yet incomplete knowledge of the cosmos. But on closer scrutiny, this novel string theoretical imaginary betrays, like the exposition within string theory technical discourse, its own incoherences. A radical heterogeneity arises from the way popularizations blend a multitude of string and brane images with what I called a *commonsense epistemological imaginary*. The imbrication of string theory images within this imaginary allows these seemingly counterintuitive images to take on a certain recognizability and legitimacy. It is the situating of these ostensibly novel string and brane images within a particular epistemological imaginary that forms a string theory imaginary as I have previously defined it.

This chapter concerns itself with yet another discourse that takes up string theory as a scientific imaginary. I refer to literature, which for the purposes of this investigation, I define as that discourse where the imagination is understood to be primary, or in other words, imaginative writing. Accordingly, this chapter confines itself to a selection of texts—including fiction, poetry, and drama—that is representative of the adaptation of string theory images in imaginative writing. The question of principal import then will be: what happens to a string theory imaginary when it is translated from a discourse that claims epistemological authority into an avowedly imaginative domain—literature?

In *The Philosophical Imaginary*, Michèle Le Doeuff offers several possible explanations as to why texts within an epistemological tradition borrow certain key images from authoritative sources.

> One can imagine a plurality of functions for this borrowing: the mask of a lack or a transgression, the recurrence of a past blind spot, itself already given in an image (when an image becomes classic one no longer sees that its reason was to soften an aporia); the repetition of a wish or desire . . . a process or attempt at validating an enterprise by placing it under the "patronage" of an already recognized authority. (92)

With respect to string theory technical discourse, it is mathematical argument that constitutes what Le Doeuff calls the *blind spot*—that locus in the text that indicates a lacuna or lack—a self-consistent, yet entirely abstract, disembodied assemblage of symbols that calls for a "physical interpretation," to quote the theorists. The aporia with which string theory technical exposition must contend lies in bringing ordinary language to bear on complex mathematical formalisms. Accordingly, within technical discourse, the expository prose that surrounds mathematical argument as a supposedly extraneous supplement works to provide imagery that doubles for, and consequently, obscures that hermeneutic blind spot. A complex amalgam of mathematical formalisms becomes a relatively incoherent heterogeneity of imagery that, in a simplifying move, culminates in what is presented as an imaginatively coherent physical "phenomenon," the string.

When considering the migration of this image of the string from technical exposition to popularizations to literary discourse, in keeping with Le Doeuff, one can view it as a case of one discourse borrowing from another an image it "cannot produce itself" (92).[1] This is because the borrowing discourse lacks the authority, within its own discursive domain, to make any claims to a particular truth-value, one that depends exclusively on mathematical rigor. Le Doeuff further asserts that an image, "far from being a more or less pedagogical 'illustration' of an abstract thesis contained elsewhere in the system, is always the mark of a tension, a signification incompatible with the rest of the work" (92). As I argued in the previous chapter, within popularizations, the image of the string or brane functions in just the way that Le Doeuff here describes—more than merely a pedagogic supplement, these images engender a "mark of a tension" or a "signification incompatible with the rest of the [discourse]." The image of the string, doubling for a "blind spot" within technical argument, comes to embody that which is absent from or incompatible with mathematical formalism, but nevertheless is necessary for string theory's comprehensibility, its imaginative adoption by a given culture.

In effect, I hope to provide a comparable etiology for literary adaptations of string theory in terms of the three "functions" that Le Doeuff claims motivate the borrowing of an image by one discourse from another. The question initially posed concerning literary discourse now becomes the more pointed task of assessing the extent to which a string theory imaginary as represented by a given literary text fulfills one or more of these three functions. Furthermore, Le Doeuff's posited functions are useful in that they also provide a framework for addressing the relationship between an image as a "physical interpretation," to quote Gross et at. again, and an epistemological imaginary as its context. The embedding of an image of the string within a given fictive, poetic, or dramatic text implies an epistemological imaginary. An imagined means of encounter with or access to those images becomes an imagined means of knowing the cosmos, and thus, the formulation of a certain intelligibility to the cosmos, a cosmic order.

My intention in this chapter is not to produce an exhaustive survey of instances of string theory's appearance in contemporary Anglo-American literature. Rather, I have chosen texts that represent typical instances of the kind of borrowing that betrays, in an unambiguous way, Le Doeuff's functions. Most of these texts fall under the purview of what is often called *genre writing*. This is writing that, while categorically imaginative as I defined it earlier, nevertheless relies heavily on the repetition of conventions.[2] As genre writing, not only do these texts tend to adhere to the stylistic and formal conventions of the genre to which they belong, they also tend to integrate string theory ideas and images in relatively unsophisticated ways. In effect, one can characterize the engagement with string theory that these texts undertake by the extent to which they reproduce the commonsense epistemological conceits that link basic string theory images to a broader imagining of a given cosmic order. As such, the first half of this chapter concerns itself with an examination of such "readerly" texts as examples of the ways in which Anglo-American popular culture consumes string theory as a scientific imaginary. The analysis endeavors to assess whether the literary texts validate the authenticity of their own "enterprise" through the "patronage" of string theory as an authority; repeat "a wish or desire" regarding what they understand string theory to be as both scientific knowledge and precursor to technological power; and radically juxtapose, as a kind of imaginative parataxis, string theory as a cosmic order to a domestic space in order to validate ethical concerns relevant to the human-scaled interpersonal world.[3] As with syntactical parataxis, the capacity for the reader to fill in the associative gaps between the juxtaposed images depends on a certain cultural literacy, a certain common-sense.

The second half of this chapter takes up two imaginative texts that offer the promise of an engagement with string theory that exceeds these relatively unambiguous functions of borrowing. The first is a play by Carole Buggé called *Strings* and the second, a poem by Brenda Hillman entitled "String Theory Sutra." This last text, in particular, is an example of imaginative writing that realizes a certain "writerly" potential, as Roland Barthes would have it, with respect to string theory as an imaginary—one that begins to achieve a sophistication, relative to its own epistemological domain, comparable to string theory technical exposition.

Borrowing Procedures

Le Doeuff's functions address the important question of why one text would borrow key images from another. It is also worth considering the question of how the borrowing takes place. In *Science on Stage*, Kirsten Shepherd-Barr surveys representations of scientific theories specifically within what have come recently to be classified as "science plays."[4] She suggests that such science plays "share certain critical features," one of which being "a direct engagement with 'real' scientific ideas" (2). This notion of string theory "ideas" appearing within an avowedly imaginative discourse becomes complicated by the issue of their source. As we shall see, in most if not all cases, it is string theory popularizations that are the source, rather than technical discourse.

The play *String Fever* by Jacquelyn Reingold is a typical example of how imaginative writing borrows explicitly from a particular string theory popularization.[5] *String Fever* centers on the courtship between a middle-aged teacher named Lily and a physicist and single father named Frank, whose daughter is in Lily's class.[6] In a key passage, Frank explains string theory to Lily and her friend Janey.

> FRANK. You really want to hear this? (*Janey and Lily nod.*) There are two theories that are the foundation for modern physics: general relativity, thanks to Einstein, which is a framework for the largest scale: galaxies, stars and whatnot, and then there's quantum mechanics, which is for the smallest: atoms, molecules, subatomic particles. They're both effective and accurate. *But,* they conflict with each other, so for decades physicists worked on one, not the other. Which was fine, except occasionally you need to look at both. With the big bang, or black holes, you have to look at both. And when you do, see, things don't make sense. No sense. At all. But you can't throw one out, because they both work. Proven. So, it's a quandary.
>
> LILY. (To *Janey.*) It's based on the idea, I think, that instead of particles being the smallest bits of matter, there are tiny strings, filaments, strands, and—
>
> FRANK.—It has the potential, the strings, mathematically and actually, to bring it all together. What physicists have been looking for. And it was discovered, really, by accident and now some, like myself, think it could be a Theory of Everything.
>
> LILY. Theory of Everything? That sounds well. Like a lot.
>
> FRANK. Yes. A lot and a little, all together. The big and the small, the utterly incongruous, disparate, incompatible—. (13, emphasis in original).

The gist of this explanation is lifted directly from *The Elegant Universe,* one particularly telling clue being Lily's use of the term "filament" to describe the string.[7]

Here the appearance of a string theory image constitutes an adaptation of an adaptation, an image that is doubly derived.[8] Accordingly, these doubly derived string theory images obtain their legitimacy from the authority of popularized string theory—as a form of received knowledge of the world. As discussed in chapter 2, for a lay audience, scientific knowledge can only ever be received knowledge—indirect knowledge mediated by the authority of a specialist expert. Such knowledge thus consists of images that are sanctioned by the string theorist as an authority to possess truth-value. String theory as a sanctioned form of received knowledge posits the cosmos as an intelligibly ordered space, which an authoritative theorist and his or her readership, as a provisional community, may imagine themselves understanding and thus inhabiting. The "reality" of "ideas" adapted from string theory as science speaks directly to the authority of that science as an imaginary to represent a cosmic order that competes to supersede the others that make claim to determining reality.

In *Beamtimes and Lifetimes*, Sharon Traweek reminds us that physicists in general see "their own profession as the revelation and custody of fundamental truth" (2). This takes on an added dimensionality with string theory popularizations such as *The Elegant Universe*, where theorists have highlighted enthusiastically its status as the "leading candidate" for a "theory of everything." In the introduction, I stressed that string theorists themselves are quick to qualify that, by "everything," they mean something specific to the technical problems that define string theory as a discipline—the mathematical reconciliation of quantum theory with general relativity. But unmoored from its formal expression by two successive mediations—technical exposition and then popularizations—it is understandable how what is meant by "everything" can balloon to take on a resonance far beyond its original professional discursive context. What was initially an expression coined by an exclusive coterie to describe an especially intractable yet worthwhile theoretical problem—in part, self-aggrandizing internal marketing—lends itself readily to an "everything" that engenders something altogether more coherent and total, charged as it is with the mythopoeic emotional appeal of the cosmic. A "theory of everything" within a play such as *String Fever*, with its emphasis on the exploration of interpersonal relationships—here, of lonely middle-aged singles struggling with accumulated emotional "baggage" and the precariousness of intimate connection—speaks to a more broadly connoted understanding of the cosmos and "our place" within it, the imaginative parataxis I alluded to earlier.

Shepherd-Barr offers *Humble Boy* by Charlotte Jones as an example of the arbitrary "scattering" of "scientific explanation" characteristic of many science plays (82).[9] She observes that the protagonist, Felix Humble, is a string theorist and yet string theory images only arise briefly in several instances of introspective speechmaking or asides (82). The play takes place in the "pretty country garden" beside Felix's parents' house somewhere in middle England. Felix's father has recently died, and he struggles to reconcile ambivalent feelings that linger from childhood with his uninhibited mother's desire to find a new love interest (1). In the pivotal third scene of the first act, we learn the extent of Felix's unhappiness. Jim the gardener discovers Felix in the act of trying to hang himself with a garden hose.

> FELIX. I'm sorry . . . I was just—I was experimenting . . .
> *Pause.*
> I often use a garden hose. As an analogy, I mean.

JIM. Oh yes?

FELIX. Yes. Yes. With superstring theory there need to be six or seven extra dimensions. We can't see them but it's like with a garden hose. If you stretch it out between two posts in a field and then you walk half a mile away and look back, it just looks like a one-dimensional line.

JIM. I'll take that off you, shall I? (Jim takes the hose off him and starts to wind it up again.)

FELIX. Yes. Yes. But if you look at the hose through binoculars, if you magnify it, a second dimension—one that is in the shape of a circle curled around the hose—becomes visible. So in the same way there could be extra dimensions in space but you can't see them because they're small and curled up, furled around one another. You see? (25)

The use of a garden hose "analogy" to illustrate how an extra dimension in string theory might be invisible to the eye comes directly from Greene's *The Elegant Universe;* to wit:

> This suggestion that our universe might have more than three spatial dimensions may well sound fatuous, bizarre, or mystical. In reality, though, it is concrete and thoroughly plausible. To see this, it's easiest to shift our sights temporarily from the whole universe and think about a more familiar object, such as a long, thin garden hose. . . . The upshot is that from a quarter of a mile away, a long piece of garden hose appears to be a one-dimensional object. (186)

Greene's quarter of a mile becomes half a mile; his "one-dimensional object" becomes a "one-dimensional line."[10] For Greene, the analogy serves an obvious pedagogic purpose: it concretizes an idea—extra hidden dimensions of space—that he suspects his lay readership may find "fatuous, bizarre, mystical." One can understand the slight distortions in which *Humble Boy* indulges as an attempt to contextualize the string theory imagery within a decidedly domestic space that emphasizes informal conversation, a dialogue conducted face-to-face, over the more formal pedagogic exposition of a popularization. This imaginative reformulation of the popularization's account is meant to capture the "essence" of the string theory image while making it serve as a concrete vehicle for the expression of an anguished inner state. The "mystical" extra dimension becomes the rather more banal garden hose as noose—and cry for help.

Felix explains to Jim that he is "working on M-theory—trying to unify the various strands of superstring theory" (30). He goes on to summarize string theory to Jim, using descriptive language that once again seems clearly to originate in *The Elegant Universe*.

> FELIX. At the root of everything we believe, I believe—a billionth of a billionth of a billionth of the size of an atom, so many noughts it would dazzle you, the perfect Planck length—there is a loop or a filament of energy—what we call a string—which is the fundamental building block of the universe. And these strings are stretched like the strings on a violin and they're vibrating to and fro. (30–31)

The term *filament* again provides a clue to this closely paraphrased adaptation, as well as the specific expression "vibrating to and fro."[11] Felix confesses that "I can hear the little vibrating strings inside my head. Even though I can't prove absolutely that they're there, I can hear the patterns they're making, like they're ringing in my ears" (31); to which Jim immediately responds, "The music of the spheres." This is an association that Greene explicitly makes in the passage in *The Elegant Universe* where he elaborates on a description of the string: "From the ancient Pythagorean 'music of the spheres' to the 'harmonies of nature' that have guided inquiry through the ages, we have collectively sought the song of nature in the gentle wanderings of celestial bodies and the riotous fulminations of subatomic particles" (135).

I will point out other probable sources of borrowing when those images arise in the close readings to come. Suffice it to say, though, that I have yet to encounter a literary text that seems to have adapted string theory ideas directly from technical exposition. The practice of adapting exclusively from popularizations seems to constitute a convention in and of itself. This prompts the question of why it has never occurred to the authors of these texts to seek out more technically rigorous sources to further validate the authenticity of their evocation of string theory. The simplest explanation is that the popularizations themselves are taken to be sufficiently authoritative as to obviate further sourcing. Those non-specialists who do attempt to decipher technical discourse may surrender readily to their perplexity knowing that the "routine boundary work" described by Felicity Mellor that separates the technical from the popular makes their half-hearted efforts the norm—a norm sanctioned by a culture where transmission from the one discourse to the other reinforces those barriers as much as it demolishes them. As we shall see, even in the case of so-called "hard" science fiction, an emphasis on concerns other than technical authenticity more than justify the ready use of string theory popularizations for scientific content.

Borrowing as Patronage

Perhaps the most expedient purpose for the borrowing of string theory imagery by imaginative texts is to exploit the reference as a form of patronage—to borrow, in effect, the authority and mystique of string theory with respect to a cosmic order. One particularly gratuitous example of this is a 1994 novel by renowned British science fiction author Stephen Baxter, *Ring: Riding the Superstrings to the Edge of the Universe*. Though the subtitle makes specific mention of "superstrings," there is no actual mention of superstrings in the novel itself, but rather, cosmic strings—a concept originating not in string theory, but in work in theoretical astrophysics during the 1980s.[12] Considering the publication date—shortly after the rise in public awareness of string theory due to the early 1990s success of string theory popularizations—most notably, Michio Kaku's *Hyperspace*—the use of "superstrings" in the subtitle is most likely nothing more than an appropriation of a fashionable scientific buzzword to pique bookstore browsers' interest; a ploy that clearly depends on the conviction that even avid consumers of science fiction would not make much fuss over the subtle but important distinction between cosmic strings and superstrings. The use of the word "superstrings" in *Ring* represents the pirating of an image from string theory that in no way engages with that image's inferences, whether basic or epistemological, yet seeks to appropriate the cachet of that image.

The American sitcom *The Big Bang Theory* represents another example of string theory adaptation in order to exploit its patronage. The show began airing on CBS in the fall of 2007 and is, as of this writing, in its fifth "hit" season. It dramatizes the relationship between two science "geeks" named Sheldon and Leonard—who share an apartment in Pasadena, California, and work at the California Institute of Technology—and a "normal" girl named Penny, a waitress-cum-aspiring actress who lives across the hall. The show makes occasional reference to the fact that Sheldon is a leading string theorist; Leonard, by contrast, is an experimentalist.[13] Every few episodes or so, in the course of a bit of comic repartee, Sheldon spouts a semi-coherent fragment of string theory jargon. In addition, the sitcom's main set—the two friends' living room—features a whiteboard on which show consultant David Saltzberg, Professor of Physics and Astronomy at UCLA, writes an ongoing theoretical physics problem that serves as a prop and as such, a source of gleeful insider recognition by the few audience members who care to take notice.[14] Clearly, *The Big Bang Theory* uses references to string theory as a means of authenticating the vocations and, concomitantly, the subculture of the main characters. As such, whatever imaginative resonances such references have are minimal. As authentication, they are another trivial example of patronage, one that only peripherally "validate[s] an enterprise," as Le Doeuff puts it, in that it gives the sitcom's characterization of the geek, a pop-cultural persona with a great deal of contemporary currency, a certain added intelligibility. The figure of the geek becomes intelligible precisely through his enunciation of techno-scientific gibberish, which is recognizably incomprehensible to the self-identified non-geek. The references to string theory also can be said to repeat a certain desire toward string theory as professional praxis, to its valorization as a vocation both exclusive and arcane.[15]

Parataxis of the Cosmic and the Domestic

As previously discussed, the embedding of an image of the string within a given fictive, poetic, or dramatic text implies an epistemological imaginary—an imagined means of encounter with or access to those images becomes an imagined means of knowing the cosmos, and thus the formulation of a certain intelligibility to it, a cosmic order. In many respects, to consider string theory a source of "real scientific ideas" is to engage, whether conscientiously or naively, with how a basic image—in particular, the string or brane, is embedded within an epistemological imaginary that surrounds the image and provides such intelligibility. It is precisely this imagining the means by which we may know the cosmos as constituted by strings and branes that allows an interplay between, to quote Michel Serres, "relations among ourselves" and a "rapport with things" to gain traction (*Conversations* 141). Perhaps to overstate the obvious, for a lay audience to accept the reality of strings and branes depends largely on the extent to which the epistemological imaginary in which these images are embedded coincides with a certain common-sense. By definition, this common-sense consists of an imaginary adopted as a form of received knowledge, one that both enjoys a certain cultural currency and is taken as a matter of convention. In the context of theoretical physics, it offers a satisfying representation of the cosmos, and particularly, an imagined means of accounting for the cosmos. It does so by presenting a cluster of basic images that are meant to represent the macro- and microcosmically remote within the context of a supplemental group of what I have called *conceits*. These epistemological conceits serve to facilitate an engagement with the images-as-cosmos. They represent widely shared and deeply conventionalized ways of structuring our understanding of the world through a limited array of models, images, and ideas. I call them *conceits* in order to emphasize their figurative quality in that they tend to ground abstraction in the relatively concrete. For example, chapter 2 discussed the imaginative distinction between heliocentrism and geocentrism. In spite of superficial evidence to the contrary, we now understand heliocentrism, as a form of received knowledge, to be the "deeper truth." The epistemological conceit of depth and surface organizes these two images, Earth revolving around sun privileged over sun moving about Earth, within the imaginary.

Taken in aggregate, a given amalgam of conceits becomes an account of a certain cosmic order. For string theory technical exposition and its popularizations, the account takes the following form: we can know the order of the cosmos as an intelligible whole. It is intelligible in that the universe consists of general kinds of things—strings and branes. These kinds of things are knowable in their essences, by attributes that define their "natural" behavior. The essence of a string is vibrating energy. That behavior, vibration, constitutes an allotropy—one substance, the vibrating string, begets many forms, the quanta, or subatomic particles. Kinds, essences, and allotropy—these are the three fundamental and highly abstracted conceits that begin to allow us to imagine ourselves in relation to a world made up of strings.

String theory popularizations extend and supplement these fundamental imaginative conceits from technical exposition. In technical exposition, strings emerge from a process of construction that allows theorists to recognize them as something natural. Strings are initially constructed but ultimately autopoietic. As a complement, popularizations imagine string theory as a depth below a surface; a deeper truth that must be found out.[16] In turn, the cosmos, via string theory, is a text that can be deciphered. Furthermore, as autopoietic objects, strings are imagined as non-human agents; that is, Planck and atomic scale events are imagined as actions. To imagine them as actions implies a certain agency—a doer that acts. By imagining strings as agents, one may then formulate string theory as an encounter between two agents—one human, the other abstractly animate. Imagining strings as exhibiting autonomous behavior further facilitates the epistemological conceit of discovery. Within remote abstract theoretical spaces, the theorist may encounter and thus discover strings, rather than simply invent them.

Popularizations further concretize these four key conceits—construct, depth, text, and encounter—implicit in string theory technical exposition. They also add another imaginative layer to this amalgam in an effort to further substantiate strings and present string theory as a coherent whole.

For popularizations, as a matter of convention, string theory represents a mystery to be solved; a revelation of something hidden (which jibes with the image of string theory as exploring a depth below a surface); a journey, whether a quest outward in search of something or a homecoming, a return to a state of knowledge; and the domestication of a wilderness-like space. These supplemental epistemological conceits further enable the radical juxtaposition of the remote with the near-at-hand. They help make the alien graspable, knowable. Any engagement with string theory on the part of literary discourse must contend necessarily with the scaffolding that these epistemological conceits supply.

Like *Ring*, a trilogy of novels—*Star Trek Voyager String Theory*—lures potential readers by co-opting the fashionable mystique of string theory. They are based on the television series *Star Trek Voyager*, whose U.S. broadcast ran from 1995 to 2001, and bear the individual titles *Cohesion*, *Fusion*, and *Evolution*, respectively. Oddly, the first mention of string theory occurs only in the second book, *Fusion*. Here an alien race called the *Nacene* explains to Janeway, the captain of the starship *Voyager*, the fundamental nature of the universe—that it is composed of strings. But it is not until the last installment of the trilogy that string theory gets any further explication. There, a quasi-omniscient superbeing named *Q*—familiar to *Star Trek* fans—explains to another *Voyager* crew member the history of the Nacene's knowledge of strings:

> "Like humans, the Nacene went out and explored their dimension—they call it Exosia—and in the process they discovered a building block even smaller than the atom. You call them strings. . . . The problem with their failure to keep their proverbial hands to themselves was that their happy-go-lucky carelessness with the strings affected more than just Exosia and your dimension, but every dimension that emerged from the Big Bang. You see, what you humans call 'strings' is a fundamental building block of the universe. All the dimensions are interconnected in a cosmic ecology that expands infinitely in all directions, and one of the key elements connecting all these pieces together are what you perceive to be "strings." The strings have different names, manifestations, and roles depending on which dimension is being considered—"
>
> "So the Nacene messed with the strings and everything got all knotted up," Tom said. (74, 77)

Here the image of the string comes couched in the terms of the conventional conceit of construction—the "fundamental building block of the universe." Furthermore, the multidimensionality of string theory is flattened, so to speak, into a figurative geography, where "dimension" signifies rather literally an isolated territory. This image is consistent with an earth-like scale where separate regions are linked in an "ecology." The qualifier *cosmic* presumably is meant to give that ecology a certain imaginative scope and grandeur.

That this imaginary is framed by a dialogue between two speakers—one human, the other highly human-like—engenders a domestic space, in the sense that this is a space where a face-to-face encounter comes to the imaginative fore. It exemplifies a pattern characteristic of much if not all of the imaginative writing presented here, and as such, a departure from the narrative structure of popularizations, where, in a pedagogic stance, the narrator addresses an implied reader. In the *Star Trek Voyager* novels, the audience for the lesson on string theory is not the reader per se, but another character. In effect, the reader is meant to bear witness to the exchange rather than participate implicitly in it. The intimacy between the characters is enhanced in that the reader is assumed to be absent, only a silent, invisible witness. The reader eavesdrops on a cloistered space that foregrounds the face-to-face encounter. The effect is akin to that of parataxis: images of microcosm and macrocosm, radically remote in scale, are framed by dialogue between characters ensconced within a relatively cloistered space—in this case, a spaceship's deck. In other texts that domestic space becomes a living room, a café, an office, etc.

The short story "On the Brane" by Gregory Benford constitutes a representative example of a "hard" science-fiction treatment of string theory that also juxtaposes the cosmic and the quasi-domestic. In his introduction to *The Ascent of Wonder*, David Hartwell defines *hard science fiction*, which evokes the distinction between the so-called "hard" and "soft" sciences, as science fiction that describes and confronts "what is scientifically true" (1). Hartwell claims that for "experienced readers," hard science fiction "feels authentic" because "the way things work in the story is scientifically plausible" (1). By definition, hard science fiction takes as one of its principal aims the maintenance of a fealty to the science that is its subject matter. As such, other notable features of fiction, in particular, complex characterization—the thorough exploration of psychological interiority—is more often than not a secondary consideration. The subtlety of hard science fiction resides in the integration of supposedly faithfully rendered scientific ideas within a reimagined universe. To put it in terms of what Darko Suvin famously calls, "cognitive estrangement," in hard science fiction, there is a special emphasis on the "cognitive," of grounding a concern with the "domestication of the amazing" in a rigorously "rational" extrapolation from the empirical

(373, 375). Hard science fiction attempts to distinguish itself from science fiction in general through its scientific accuracy. This implies, in a certain respect, a close alignment with the pedagogical authority, and thus rhetorical strategies, of popularizations.

"On the Brane" was chosen for inclusion in a prominent anthology of science fiction entitled *Year's Best SF*, edited by Hartwell and Kathryn Cramer (310–29). Benford himself is well-known in the hard science fiction galaxy. In addition to winning the Nebula Award for Science Fiction twice (in 1975 and 1980), he received the United Nations Medal in Literature in 1989 and an Isaac Asimov Memorial Award in 2007. As a professor of physics at the University of California, Irvine, Benford is in a particularly advantageous position to represent scientific knowledge authoritatively, though it is important to note that he is not a string theorist, but an experimental cosmologist.[17]

In the story, two astronauts, Mina and Ben, pilot a spacecraft across a hidden, extra dimension to a dying world called "Counter-Earth" that parallels our own. There they encounter primitive animal-like creatures who are the survivors of a long-extinct advanced civilization. Ironically, the way Benford presents the "hard" scientific content that underpins the central premise of the story suggests an epistemological imaginary that, once again, depends more on a form of patronage, a submission to an authority, than an accurate rendering of string theory as science:

> Mina had only a cartoon-like understanding of how another universe could live on a brane only twenty centimeters away from the universe humans knew. The trick was that those twenty centimeters lay along a dimension termed the Q-coordinate. Ordinary forces couldn't leave the brane humans called the universe, or this brane. But gravity could. So when the first big gravitational wave detectors picked up coherent signals from "nearby"—twenty centimeters away!—it was just too tempting to the physics guys. And once they opened the portal into the looking-glass-like Counter system—she had no idea how, except that it involved lots of magnets—somebody had to go and look. (313)

"On the Brane" asks the reader to "intuit" Counter-Earth by playfully dismissing the need for Mina to comprehend the actual science of branes. Nevertheless, apart from this omission, the story indiscriminately proffers a cornucopia of terms and images adapted from a myriad of other scientific disciplines, including chemistry, biology, anthropology, and linguistics. Yet in this one crucial sequence, on which the whole story hinges, the narrator is content to withdraw from the usual precepts of "hard" science fiction, which generally demand a certain rigor in the representation of that which is marked as scientifically authentic.

At the same time, "On the Brane" effects a systematic concretization of the images that permeate string theory popularizations; for example, the image of the space-time "slice" and the multidimensional "fold." The use of such images by "On the Brane" is both an acknowledgement of the highly specialized nature of a scientific practice, where string theory as a discipline both suffers from and exploits certain epistemological boundaries, and an arrogation within the story itself of the prerogative of transgressing that boundary through an imaginative leap of faith. This is a double move that, on the one hand, denies access to or obscures the specialized knowledge of string theory, and on the other, by the authority of the narrator, replaces that knowledge with a conventionalized imaginary of the "looking-glass," as it were, a recombinant world just beyond our own.[18]

As in string theory popularizations, a story such as "On the Brane" functions as yet another form of "routine boundary work," to revisit Mellor's argument in "Between Fact and Fiction." Mina comes to recognize the fate of these alien creatures in a world with a weakened gravity as a possible human future, and as such, the story's use of the brane image is ultimately cautionary. Nevertheless, it substantiates string theory by imagining its astronauts journeying to another dimension where they initiate a domestication of a wilderness-like place. In a certain sense, "On the Brane" constitutes a concretization of the abstracted encounter of string theory technical exposition discussed in chapter 3. The fantasy of an encounter with aliens on a not-so-strange Counter-Earth, a world only counterintuitive in a relatively trivial way, in effect, serves to perpetuate a somewhat puerile wish regarding string theory's supposed power.[19] "On the Brane" does so by privileging an epistemological imaginary where a mystery is solved (what happened to Counter-Earth) and an essence is laid bare (the affinity between two intelligent life forms)—and by glossing over the possibility of the radical incomprehensibility of the alien and remote.

Another example of the parataxis of the cosmic and the domestic is the previously discussed play *String Fever*. *String Fever*, like other science plays, exhibits a certain "visual minimalism and textual abundance," as Shepherd-Barr puts it, a "scenic restraint" that "de-emphasize[s] spectacle" and serves to "foreground both the text and the actors' bodies" (2). When it is effective, this visual minimalism is meant to (literally) put a spotlight on a stark space of scientific exposition. The character Frank, as a string theorist, possesses a mystique that is predicated on his authority to posit an "answer" to the great cosmic mystery. In a rather trite double-duty, this authority in turn suggests that such a persona might promise to be the "answer" to Lily's loneliness, to her chronic romantic uncertainty. Much as string theory reconciles the "incongruous, disparate, incompatible" into one neat formalism as a "theory of everything," in the logic of the play, a romantic relationship with Frank offers a parallel possibility. String theory as cosmic order suggests romantic fulfillment within a domestic space. But again, this coupling of strings with the "fever" of romantic love seems gratuitous, an arbitrary juxtaposition of microcosmic images and affairs of the heart. The play's principal concern is an exploration of Lily's emotional isolation, with little effort made to complicate her and the audience's understanding of the interpersonal relationships in which she finds herself embroiled specifically through the prism of string theory as an imaginary.

From Science to Technology

In the latter part of the previous chapter, I described the ways in which popularizers—Kaku and Greene, in particular—extend their expositions of string theory as science to a discussion of what Le Doeuff would characterize as a wish or desire regarding string theory as science. In *Hyperspace* and *The Elegant Universe*, the two authors speculate on the potential practical applications of string theory, of transforming the science into technology.[20] Implicit in these speculative flights is the equation of string theory with power—firstly scientific knowledge as one manifestation of an abstract power that Kaku couches in quasi-religious terms: "being blessed with the intellect to divine the ultimate secrets of nature gives meaning enough to life" (334).[21] For Kaku and Greene, though, the technological innovation that results from scientific knowledge is the more readily comprehensible, and perhaps attractive, manifestation of that power. They both want to understand the promise of string theory as a potential form of "mastery" over the cosmos. Greene writes that string theory may one day yield a "master equation" and Kaku that "anyone, or any civilization, that truly masters the energy found at the Planck length [10^{-33} centimeters] will become the master of all fundamental forces" (269). Technological applications that result from such mastery might include the exploitation of "wormholes" to travel back and forth through time, as well as into and amongst extra dimensions. In keeping with this desire, Kaku speculates on the possibilities of contact with a "sister universe," akin perhaps to Benford's "Counter-Earth."[22] He goes on to suggest that a civilization with "mastery of the tenth dimension" eventually might have "the energy output of uncountable star systems and perhaps the galactic nucleus at its disposal" (280).

Almost as a matter of dogma, Kaku and Greene reinforce the stock epistemological conceit of knowledge as depth coupled with technological progress as a journey leading from past to future that results in the increasing capacity to "harness" nature, and thus to control it. Greene concisely summarizes this grand narrative of historical mastery in *The Fabric of the Cosmos.*

> While there is no set pattern to scientific discovery, history shows that deep understanding is often the first step toward technological control. Understanding of the electromagnetic force in the 1880s ultimately led to the telegraph, radio, and television. With that knowledge, in conjunction with subsequent understanding of quantum mechanics, we were able to develop computers, lasers, and electronic gadgets too numerous to mention. Understanding of the nuclear forces led to dangerous mastery over the most powerful weapons the world has ever known, and to the development of technologies that might one day meet all the world's energy needs with nothing but vats of salt water. Could our ever deepening understanding of space and time be the first step in a similar pattern of discovery and technological development? Will we one day be masters of space and time and do things that for now are only part of science fiction? (436)

For the most part, Randall echoes these same epistemological conceits in *Warped Passages.* She also characterizes the universe as "wonderful and mysterious" and asserts that "our goal is to learn how its pieces fit together and how they've evolved into their current state" (456). This goal involves the continuing effort to "understand the ultimate origin of matter at the deepest level," an activity she sees as a form of puzzle-solving (456). Elaborating on the wonder of the cosmos, Randall alludes to the epistemological conceit of the wilderness in the way she describes extra dimensions. She writes that "extra dimensions have opened our eyes and our imaginations to amazing new possibilities. . . . The cosmos could be larger, richer, and more varied than anything we imagined" (456). In effect, to "pry open" or "solve" the cosmos is, in some sense, to domesticate it, to imaginatively contain the radical variety of it. Emphasizing the uncertainty theorists currently face, she writes:

> Despite the impressive physics developments of the last few years, we don't yet know how to harness the force of gravity or teleport objects

> across space, and it's probably too soon to invest in property in extra dimensions. And because we don't know how to connect universes in which one could loop through time to the one in which we live, no one can create a time machine, and most likely no one will do so any time soon (or in the past). (455)

The subtle difference between Randall's take and that of Greene and Kaku is that Randall tends to privilege the value of the scientific knowledge itself as its own end over its application through technology as a display of mastery.

Perhaps picking up on the concerns of the popularizers, many of the literary texts that evoke string theory similarly are preoccupied with it as a form of scientific knowledge that offers the prospect of technological mastery. Like Kaku's *Hyperspace*, *Star Trek Voyager String Theory* conflates the image of the string with that of an ecology. *Evolution* compounds an imagining of the cosmos as organic, animate, living, with one where the cosmos is primarily inanimate, mechanistic, and coldly material.[23] On the one hand, the cosmos is something that has a life of its own; on the other, it possesses a mechanical structure that may be understood and thus manipulated. In imagining the cosmos in this way, *Star Trek Voyager String Theory* repeats the wish or desire of the popularizers regarding string theory as science: the discovery of strings as fundamental to the very structure of the universe ultimately will afford the discoverer a certain technological power. Greene asks whether one day "we" will be "masters of space and time and do things that for now are only part of science fiction" (436). Concomitantly, *Star Trek Voyager String Theory* imaginatively concretizes that mastery; the two texts, while ostensibly belonging to distinct discourses, buttress a shared epistemological imaginary.

The twist, though, is that in the imaginative text this desire for power is presented as part of a cautionary tale that echoes the current anxiety in Anglo-American culture concerning the deleterious impact of human industry on the ecology of the planet—ranging from the terror of catastrophic nuclear holocaust to the gradually mounting anxiety of global warming. In some sense, the presentation of the "Nacene" exploitation of strings serves as a tacit rebuttal of the optimism implicit in Kaku and Greene's equation of string theory with the power of a monolithic human civilization. In spite of this twist, *Star Trek Voyager String Theory* is neither a very substantive engagement with the scientific ideas of string theory nor a very complex ethical discussion. And since it is a piece of genre fiction, the conventionality of the prose itself does not lend itself to an exploration of string theory as an imaginary through an experimentation with novelistic form. The use of string theory imagery in these *Star Trek Voyager* novels suggests a pattern that is borne out in numerous other fictional texts that, like the *Star Trek* texts, do little more than evoke the string or brane as a vehicle for an appeal to caution, while nonetheless romanticizing string theory as scientific knowledge through a concretely realized technological scenario. While promoting the need for caution, that act of imaginative concretion itself unambiguously repeats the popularizer's wish concerning the future, structured, as discussed in the previous chapter, as a romance of power. As such, it implicitly endorses the epistemological imaginary that sustains that desire.

Another prime example of this equation of scientific knowledge with technological power with respect to string theory is Mark Alpert's novel *Final Theory*. Trained as a physicist, Alpert is currently an editor at *Scientific American*. As a thriller in the tradition of Michael Crichton or Dan Brown, *Final Theory* asserts a certain popular scientific authority within the framework of the conventional thriller plot, where an outsider hero attempts to solve a mystery while running the gauntlet of a powerful conspiracy, using an amateur expertise in a given field to solve the mystery. In this case, that field is string theory, which the text establishes at the outset as a central theme around which the plot is organized. *Final Theory* is preoccupied with a highly coveted piece of esoteric knowledge—the fabled "unified field theory" attributed to Albert Einstein. After overcoming numerous life-threatening obstacles, the protagonist, David Swift, with the help of a physicist colleague and love interest, Monique, finally stumbles upon Einstein's long-lost theory.[24]

> "That's what the diagrams are showing. Two empty branes collide and the energy from the crash fills our universe, eventually turning into atoms and stars and galaxies, all of them hurtling outward in a gigantic wave." She grabbed his sleeve again and looked him in the eye. "This is it, David. The answer to the mystery of Creation." (278)

In a liberal use of poetic license, the narrator attributes Paul Steinhardt and Neil Turok's contemporary model of the "ekpyrotic" universe to Einstein.[25] But the result of this conflation does not serve as an aperture into a concerted exploration of Big Splat imagery. Rather (leaving aside the inadvertently comic allusion to copulation), it is reduced to an epistemological banality—the "answer" to a "mystery"—that pertains, rather perfunctorily, to all of "Creation." Further explanation on the part of Monique, which employs jargon such as "spacetime," "sterile neutrinos," and "the bulk," leads to the realization that "you could also use [the equation] as a weapon" (280). This realization in turn suddenly makes apparent to David the motivation for his antagonists' ruthless pursuit of him.

> Now David knew why the FBI had chased them halfway across the country. A weapon like this would be perfect for the war on terror. The Pentagon could eliminate its enemies without deploying commandos or cruise missiles. Because the particle beam would travel through the extra dimensions, it would evade radar, antiaircraft fire, and all other defenses. (280)

While it might be said that the novel's engagement with "real scientific ideas" is slightly more sophisticated than that of the *Star Trek* trilogy in its robustly confident pastiche of jargon drawn from string theory, this engagement fails to add anything more nuanced to the plot or characterization. It is a piece of conventional genre writing, and string theory ideas certainly do not inform the structure of the novel itself. Like *Star Trek Voyager String Theory*, *Final Theory* rather transparently repeats the desire for string theory to embody a certain notion of power consistent with its prominent popularizations. It validates that desire through the semi-coherent use of string theory jargon. In its cobbling together of string theory images and injecting them into dialogue pell-mell, the novel serves to reinforce unquestioningly the commonsense epistemological imaginary that serves to mark string theory as science in popular culture. That science affords the text a certain authority and prestige—and as a corollary, validates the promise of string theory as epistemological and ultimately technological power. The novel's only contestation of this string theoretical epistemological imaginary is to present it in a way that tepidly cautions against the abuse of the power that issues from it, while nevertheless contradictorily glorifying the "mystery" of its insight and originality.

A more sophisticated meditation on the potential consequences of string theory as techno-scientific power is Adam Roberts's short story "S-Bomb."[26] The story is framed by two secretive meetings that take place at an urban coffee shop between a young male physicist and an older male defense ministry officer, both unnamed. The physicist is part of a government-funded research team that has developed and tested an "S-bomb," a weapon that unleashes the energy within strings to destructive effect. Unlike an A-bomb, the S-bomb's impact is not explosive but rather, implosive; the detonation of an S-bomb "unweaves" the multidimensional space-time fabric, leaving a devastating "vacuum" or "null space" in the exact location in the universe where it went off (34). The first meeting occurs just after this first test. The second meeting happens six months later. The technology of the S-bomb is easy to replicate, the bomb itself easy to reproduce. In six months, S-bomb detonations have proliferated throughout the world, dangerously destabilizing the political status quo and setting off a global economic depression—as the night sky becomes littered with blotches of starless, opaque "null spaces."

"S-Bomb" attempts to explore the inferences of the basic image of the string in a way that resonates imaginatively with the story's broader ethical concerns. What follows is a key passage where the young physicist explicates the "hidden" nature of the world as made up of superstrings.

> You think superstrings are myriad little-little separate strings, one-dimensional extended objects that resonate and shake, that aggregate and disaggregate into subatomic particles, and thence into atoms and molecules and everything in this diverse and frangible world. You think so. Think again. Think *laces.* Think of it this way: one single string, ten-to-the-million meters long, weaving in and out of *our* four dimensions, like laces weaving in and out of cosmic fabric, tying it together. Superstrings is a misnomer. This singular thing, this superstring. The equations require ten dimensions, and we're personally familiar with four dimensions, and all that is true. But when you look at it clearly, there *is* only one dimension. Only the one singularity, the thread that ties all of reality together and also the thread out of which all reality is woven. The one string. (34, emphasis in original)

The young physicist here quotes verbatim the definition of the string as given by its popularizers: a "one-dimensional extended object that resonates." This is a definition that implies the standard epistemological conceits of string theory mentioned above (i.e., kinds, essences, allotropy). But then the physicist counters this imaginary that emphasizes the individual "separate"-ness of these strings by insisting that this plurality of distinct yet essentially uniform objects, "superstrings," is, in fact, a "misnomer." A reversed formulation of the image emphasizes the connectedness of superstrings, that there is "only the one singularity, the thread that ties all of reality together."[27] Here, the image of the string takes on a dual significance. This recognition of the "singularity" of superstrings as a contiguous "fabric" of reality suggests a comparable stance regarding social relations—where strings stand for humans as individuals. This revelation of a previously obscure cosmic order implies that within a human-scaled domestic space, our interconnectedness ought to be emphasized over an individualism that tends to solipsism. Ironically, this scientific realization of the universe's fundamental interconnectedness precipitates a scramble of arms proliferation that further concretizes an utter failure on the part of humanity to recognize its own mutual interconnectedness and its collective impact on a "lace"-like, "diverse and frangible" cosmic order. Such an image undermines the epistemological conceit of domestication, suggesting instead that attempts to "harness" strings might lead logically not to mastery but rather, to a disintegration of both the political and cosmic order.

While the texts presented thus far in this chapter explore, with varying degrees of sophistication, the implications of string theory as a form of techno-scientific power reminiscent of popularizations, other texts, such as short stories by Rudy Rucker and Elissa Malcohn, imagine string theory as a form of power distinct from technological exploitation.[28] Accordingly, these stories deemphasize "hard" scientific content and in its stead offer what is normally defined as *fantasy*. Shepherd-Barr observes the "tendency toward fantasy and away from 'hard' science [that] characterizes the way in which science is dealt with in other media like film and television" (8). Producers and consumers of hard science fiction often couch this move away from science (as they define it) to fantasy in disparaging terms, the tendency of less "serious" media such as television. But in the case of Rucker and Malcohn, this turn to fantasy constitutes an alternative form of boundary work.

The narrator of Rucker's short story, "Guadalupe and Hieronymus Bosch," is Glenda Gomez, "an unemployed overweight unmarried overeducated woman with a big mouth" who lives in an apartment complex in San Jose, California (90). Glenda encounters a polymorphic extraterrestrial named Harna who can travel through space and time. Harna helps Glenda abduct Hieronymus Bosch, her favorite artist, from the "Lowlands of 1475" (94). In the following passage, Glenda describes Harna's initial appearance in her apartment.

> "Who are you?"
>
> "I'm Harna from Hilbert space." She has a prim voice; I visualize flowery dresses and pillbox hats. "I happened upon your brane several—days—ago. I've been teeming with the microlife, a bit humdrum, and I thought that's all there is to see in this location. Worth documenting, but no more than that. I had no idea that only a few clicks up the size scale I'd find a gorgeous entity like you. Scale is tricky for me, what with

> everything in Hilbert space being infinite. Thank goodness I happened upon your blood cell. Oh, warmest greetings, Glenda Gomez. You're—why, you're collectible, my dear."
>
> I'm fully buggin'. I run to the corner of my living room, staring at the luminous paramecium the size of a dog in mid-air. "Go away," I say. (92–93)[29]

On the one hand, the evocation of the image of the brane here is not particularly nuanced; it suggests a place that serves the rather banal consumerist interests of Harna, a kind of cosmic flea market or mall where she hops from brane to brane shopping for "collectibles." But on the other hand, this passage epitomizes the tone and scope of the story; it is a paean to excess and heterogeneity—the infinity of "Hilbert space" comes to signify a self-conscious macaronization of American suburban culture. Rucker mixes images from a wide array of scientific discourses with abandon—branes, paramecia, blood cells, etc. This juxtaposition of images hinges on the importance of scale as a conceit in a string theory imaginary. Through its use of the scale-bending fantastic, the story implicitly rebuts the romance of string theory as techno-scientific power. The power that Harna enjoys is harnessed not to technological mastery per se, but to the ostensibly frivolous purposes of hobby collecting and matchmaking. The power of string theory as scientific knowledge functions here as no more than a kind of magical thinking that liberates the imagination from the stultifying confines of domestic loneliness.

"Arachne" by Elissa Malcohn also eschews the ostensible realism of hard science fiction for the imaginative possibilities of fantasy. The short story blends a contemporary domestic setting (the story takes place in a house) with ancient Greek mythology and string theory; it reimagines the rivalry between the goddess Athena and the mortal Arachne as weavers into a more cooperative relationship where Athena recruits Arachne to "untangle the multiplicity of Chaos when the universes met" (98). "Arachne" is a coming-of-age story that explores through an unconventional familial dynamic between mother and daughter the possibilities of artistic creativity. In a pivotal expository passage, Athene muses that the universe is much like a web.

> *We are at the perimeter of a large space*, Athene mused, *and we are expanding toward its center*. What's more, the discoverers of the ten dimensions had also included, in their Theory of Everything, the fundamental building blocks of matter and energy and named them:

> Strings.
>
> If Arachne was worthy, she, like her counterparts, would not only be the supreme weaver. She would become the supreme Creatrice. (98, emphasis in original)

By manipulating these strings, conventionally imagined as "fundamental building blocks of matter and energy," Arachne may create an ordered whole out of the universe much as a spider makes a web. Furthermore, the "ten dimensions" of string theory are imagined as a discovery, another conventional epistemological conceit. What is novel about this scene is not the conventionality of the image of string as building block, but rather its imagination of our place in the universe as not the center but the perimeter. Like a spider, Arachne's creativity works from this boundary inward toward a figurative center. Malcohn upends the commonsense cosmic order of center first and then periphery in an imaginative textual act that emulates the intricacy of the spider as it weaves a web. Through this mimicry of a natural image—the spider's weaving that moves from perimeter to center—Arachne becomes capable of exercising a certain kind of cosmic power, the power to untangle chaos, to weave together the fabric of reality. This idealization of string theory as *techne* contrasts with that of hard science fiction, where power is predicated more on the collectively technological than on individual craftsmanship. By exploiting the "natural" image of the spider web to elucidate the other "natural" image of the string, "Arachne" manages to disrupt the epistemological imaginary of string theory as both a construction *and* a discovered place in a way that seeks to privilege a more intimate, individual creativity. In some sense, a story like "Arachne" is also a form of boundary work (that may or may not be considered "routine"). Although the text reproduces the image of the string in a relatively unsophisticated way that suggests an attempt to place the discourse under string theory's epistemological "patronage," nevertheless, "Arachne" thwarts the repetition of the desire for string theory to represent a certain techno-scientific power, and in its stead, offers a valorization of artistic creativity. In effect, "Arachne" attempts to cordon off its own textuality, as literary discourse, as legitimate epistemological practice. Through a reworking of the Greek myth of Arachne, it substantiates a means of reconciling interpersonal relationships in a domestic space with a certain imagining of a cosmic order predicated on strings.

As discussed earlier, in reproducing string theory images from a particular popularization, the play *Humble Boy* unambiguously reproduces the epistemological imaginary in which those images are embedded: the notion that strings constitute a hidden essence, a fundamental building block of nature, and a mystery whose unraveling "would unveil . . . the secrets of the universe" (22). This faithful reproduction of string theory as "woolly and descriptive" knowledge of the nature of the universe gives it a certain authority, which *Humble Boy* exploits not to repeat the desire for techno-scientific power—or to caution against it—but rather to a purpose more in keeping with that of "Guadalupe and Hieronymus Bosch" and "Arachne." Felix's use of string theory as an imaginary serves to offer the hope not of the artistic liberation that emerges from the unleashing of individual creative power, but rather of a liberation from the psychological cleavage that Felix suffered as a child. The scene with Felix, the gardener Jim, and the garden hose captures a moment of unanticipated intimacy. Felix becomes a pedagogue in a clumsy attempt to make an excuse for his behavior—and to relieve his and Jim's mutual embarrassment. Jim becomes Felix's *de facto* confidant, an audience for his scientific expertise. This connection between Jim and Felix mitigates against Felix's accumulated failed connections with others.

In effect, the discovery of Felix's heretofore hidden suicidal thoughts—seemingly inconsequential from afar—takes on a certain significant breadth, a certain dimensionality, up close. As Felix goes on to explain to his ex-girlfriend Rosie: "It's like my mother was the big force—gently warping everything around her. And my father was the little force, fizzing away quietly on a microscopic level. But I can't bring them together" (44). In this analogy, Felix equates his mother with gravity, his father with quantum "jitters," which is not a force exactly, but an image of the subatomic world frequently presented in string theory popularizations. It is important to stress that the accuracy of the analogy is secondary to its evocation of affect, the ambivalence Felix feels as a consequence of his internalization of the conflicts that result from his parents' fundamental incompatibility. Like the texts previously analyzed, *Humble Boy* presents a domestic interpersonal dynamic predicated on an underdetermined string theory imaginary. Once again, a parataxis is at stake, an arbitrary juxtaposition of the remotely alien with the familiarly domestic. In this regime, string theory is meant to suggest a certain possibility for intimate reconciliation—the "deeper truth" of string theory buttresses the realization that the characters could and ought to get along. String theory marks the possibility that Felix may reconcile these disparate forces, "the little vibrating strings inside my head," and achieve a psychological cohesion, thereby renewing an intimacy with Rosie, in particular. *Humble Boy* reflects a concerted effort to employ a string theory imaginary in a novel way to elucidate interpersonal psychology—an interdependence of string theoretical expository content with a wholly domestic subject matter.

Buggé's *Strings*

As mentioned previously, Shepherd-Barr argues that science plays can be defined by their attempt to engage directly with "real" scientific ideas. She proposes two other criteria for evaluating such texts: firstly, whether they undertake "a complex ethical discussion," and secondly, whether they exploit "an interdependence of form and content . . . to convey the science" (2). As we have seen, much of the imaginative writing that takes up string theory disappoints in its relative failure to achieve either a particularly nuanced ethical discussion or any degree of interdependence between the form of the text and its string-theoretical content. The latter part of this chapter concerns itself with two texts that fulfill Shepherd-Barr's criteria more thoroughly than the texts thus far examined.

The first, which I would describe as a partial achievement, is the play *Strings*, by Carole Buggé.[30] The play takes place "on board a train from Cambridge to London" in 2002 (185). The protagonists are three scientists: June, an American cosmologist in her forties; George, a slightly older "upper class" English cosmologist; and Rory, a "lower middle class" English particle physicist in his forties (185). The three of them are travelling to London to see Michael Frayn's play *Copenhagen*. In act 1, we learn that June and George are married, but it soon emerges that June and Rory have been having an affair. This is the play's central ethical concern: a meditation on the causes and effects of marital infidelity, or, to put it in spatial terms, a love triangle. *Strings* blends its concern with infidelity with an exploration of the interpersonal circumstances out of which physicists make discoveries.

Shepherd-Barr argues that interdependence of form and content in science plays often manifests itself as "an extended theatrical metaphor" where characters "literally *enact* the idea that they engage" (6, emphasis in original). *Strings* presents such an "extended theatrical metaphor" through the form of the play itself. As mentioned earlier, June and George are married in the first act. The second act begins at the same moment in time as the first—June and Rory have boarded a train from Cambridge bound for London and wait impatiently for George, who arrives late. Yet in the second act, the male roles in the triangle are reversed: June is married to *Rory* while having an affair with *George*. The two acts together constitute a pair of parallel universes in the "space-time" of the play; the worlds are exactly alike except that the role of husband and secret lover are reversed. As such, the structure of the plot loosely mimics the structure of its scientific content. Throughout both acts, the three characters, two "cosmologists" and a "particle physicist," debate the relative merits of string theory and M-theory in their efforts to formulate a theory of everything. In the play's denouement, they arrive at a scientific epiphany based loosely on Steinhardt and Turok's model of the "ekpyrotic" universe.[31]

I want to stress a perhaps obvious point—that the connection between Steinhardt and Turok's ekpyrotic model and this conceit, where the roles of Rory and George in the love triangle are reversed in the two acts, is entirely a matter of poetic license. I make this point because it calls attention to the status of the abstract theoretical space of the play, in its loose mimicry of the ekpyrotic model, as a vehicle for the exploration of a particularly domestic space—specifically, the tensions of a love triangle played out within the confines of a train compartment. Importantly, the link between the ekpyrotic model and the love triangle (an interpersonal conflict figured geometrically) is predicated not on a steadfastly unambiguous similarity in their structures, but rather on the play's elucidation of the pedagogic space of string theory, of the authority of the popularizer to make accessible a particular imaginary such that the play's dual acts are understood as imitating that imaginary as such. The pedagogic access to the ekpyrotic model afforded the audience by the popularization allows for a conventionalization of what is recognized as the imaginary's basic structure—the collision of two "worlds"; that is, each act, within an extra dimensional space, the entire play. As a consequence of the conventionalization of this link, the social relations at the heart of the play can adopt the authority of the ekpyrotic model and concomitantly, its supposed emotional resonance.

As we have seen in the previous chapter, part of the appeal of popularizations such as *Hyperspace*, *The Elegant Universe*, and *Warped Passages* is that, in order to explain string theory as a prospective theory of everything, they take pains to contextualize it in terms of its predecessors. They endeavor to summarize the history of contemporary theoretical physics from Newton's theory of gravitation through electromagnetism, special relativity, general relativity, and finally, the more recent developments in quantum theory. This expository strategy tends to reinforce string theory's claim to being a theory of everything. There is an implicit correlation made between string theory as a specific technical solution that reconciles the standard model with gravity and its suggested historical status as the culmination of several centuries of scientific progress. Accordingly, *Strings* takes up as its subject matter the entirety of that tradition; namely, the history of contemporary physics through the lens of string theory popularization. The effect is to proffer a kind of crash course in contemporary physics that aligns the play with the pedagogic space of popularizations and works to buttress the sense of string theory offering a total, if not entirely coherent cosmic order.

In act 1, when June has gone to get a cup of tea, George unexpectedly confronts Rory about the affair: "She is rather like an electron being shared by two atoms, isn't she? So that makes the three of us an odd kind of molecule" (206). George then elaborates on the conceit.

> So there you were, a happy little hydrogen atom, just floating along minding your own business, feeling a bit thirsty perhaps, thinking of bonding with some juicy little oxygen atom, when along comes this absolutely knockout electron cloud, and you can't resist—wham! The two of you make a big, fat water molecule. Suddenly, there it is: ladies and gentlemen, chemistry in action! (206)

The core image, of the characters as subatomic particles, is muddled, suggesting only a haphazard concern for scientific accuracy. As a microcosm, the image completely lacks coherence. On the one hand, the three are proton, neutron, and electron, respectively.[32] On the other hand, in George's variation on the image, June remains an electron "cloud," but Rory becomes a hydrogen atom. Rory, the hydrogen atom, feels "thirsty" and hopes to "bond" with a "juicy little oxygen atom," a mixed metaphor on which, obviously, the barb hinges.

But then, in the image, June is equated not to an oxygen atom, but to an "electron cloud," which is a conventional image in popularizations that emphasizes the electron's dual status as wave and/or particle, the difficulty of pinpointing its precise location "around" the atomic nucleus. On the one hand, the suggestion is that June as a woman is contradictorily both atom and subatomic particle(s), two distinct levels of organization. On the other hand, the electron cloud is June's outer aspect, her appearance as an attractive female. Rory, the hydrogen atom, "bonds" with the electron cloud, taking it into his "orbit," which forms a water molecule. Such a transformation presupposes the fact that the oxygen atom is already bound to another hydrogen atom, since it takes two hydrogen atoms and one oxygen atom to form a water molecule. Presumably, then, Rory and George are the two respective hydrogen atoms that "share" the oxygen atom; together the three of them constitute a water molecule. But this is not entirely clear. In one sense, this image's lack of coherence nicely illustrates George's emotional turmoil, having just confronted Rory about the affair. But also, significantly, in shifting indiscriminately from subatomic to atomic to molecular scales, it demonstrates that either the play's principal concern is not necessarily to submit entirely to the pedagogic authority of string theory popularizations or that it does not fully grasp the rudiments of the science. What *Strings* does is to rearrange the imaginary as formulated by popularizations by exploiting its lacunae, the gaps where it too betrays its own relative incoherence. It is through these lacunae that the play may intercalate its own imagery to redefine the abstract theoretical space in a way that allows "physics"—with all its authority, albeit notional, in this instance—to substantiate a domestic space that reimagines the possibilities of marriage.

Radical heterogeneity works to resist essentialism. June asks, "aren't people as mysterious as the forces in an atom, in their own way?" (192). The familiar epistemological conceit is "mystery," so common to string theory popularizations. To return to Le Doeuff, for an epistemological regime that distinguishes between concept and image, each category orients itself differently with respect to mystery—what it marks as the unknown. Popularizations emphasize the conviction that string theory, as a scientific discourse, is aligned purely with concept. As such, in their epistemological imaginary, string theory's purpose is to solve the mystery, to know as completely as possible the unknown. Conversely, the role appropriate to imaginative discourses—and in this case the science play *Strings*—is, among other things, to exploit ambiguity, or to put it differently, to call attention to the unresolvability of mystery. Mystery is not an absence of knowledge, a gap to be filled, but rather a positive epistemological conceit, something to be desired for its own sake.

The question then becomes not whether images that are twice removed from string theory technical exposition become oversimplified or distorted in *Strings*, but how these images and the epistemological conceits that legitimize them serve to foreground the positive ambiguity of a certain interpersonal conflict within a domestic space. A case in point: sticklers for what is often referred to as *scientific accuracy* might object to the liberties *Strings* takes in its attribution of professional expertise to the respective characters. Rory is a "particle physicist" who works on M-theory; George, in contrast, is a "cosmologist" whose expertise pertains to string theory. Rory and George banter over the relative merits and flaws of each theory without seeming to recognize the close relationship between M-theory and its predecessor, string theory—that M-theory emerged as a direct response to problems in string theory. Buggé seems to have misunderstood that relationship—that a specialist in M-theory in all likelihood would be extremely familiar with string theory. But it is precisely this inaccuracy—and the relative unimportance of such inaccuracies—that provides an epistemological lacuna, a fuzziness in the conceptual status of the image of, for instance, a brane. The playfulness inherent in such perhaps *in*appropriate simplification thereby affords a means of redirecting emphasis away from the technicalities of string theory to an articulation of the interpersonal:

> GEORGE. Oh, but M-theory is so trendy just now; it's the Next Big Thing. What a charming concept: all matter sitting on these subatomic membranes floating around like giant bedsheets. And we're sort of like fleas hitching a ride, clinging on for dear life.
>
> RORY. Well, string theory is getting a bit tired, isn't it? I mean, you string theorists are all—forgive me—rather tied up in contradicting theories. (197)

On one level, this repartee between Rory and George alludes to the ongoing public debate concerning string theory's legitimacy as science, exemplified by books such as Peter Woit's *Not Even Wrong* and Lee Smolin's *The Trouble with Physics.* The dialogue simulates, however whimsically, the kinds of casual conversations that might take place amongst practicing theorists outside of professional contexts. This is a social milieu where the caprice of personal biases trumps the technical precision of cold, hard reason, play over concept. In this sense, the playful tone aligns itself with the intimidation that a non-specialist might feel in coming to terms with a scientific knowledge marked as impenetrably complex. That intimidation is deflected into a witticism that recognizes its own superciliousness. Here Rory and George speak both for and against M-theory. This is not necessarily a calculated resistance to an epistemological regime that emphasizes so-called scientific accuracy above all else, but, at the very least, an implicit indifference to such precision.

In act 2, with the role of cuckold reversed, a similar buildup to a confrontation occurs. There is a crucial difference, though. At the conclusion to act 1, June, George, and Rory direct toward the audience a litany of desires and beliefs in an increasingly rapid-fire exchange. Together their utterances converge to express a yearning "to believe," an open-ended, outwardly directed desire to "keep dancing . . . in the dark" (223). Toward the conclusion of act 2, the three of them begin to spiral not outwards toward the audience but inwards, toward each other. An intermingled debate over string theory and infidelity culminates in a creative professional collaboration. The denouement of *Strings* occurs thus:

> RORY. Maybe this is our chance to—
> JUNE. To what?
> RORY. To make it all come out right this time. (Pause. To George) Your strings have to vibrate in more than three dimensions, right?
> GEORGE. Right—in order to accommodate all the particles you particle physicists have discovered.
> RORY. What if they vibrate in eleven dimensions?
> JUNE. Like in M-theory?
> GEORGE. Why?
> RORY. To accommodate gravity.
> . . .
> JUNE. What if two of the membranes collided—sort of like you and George?
> RORY. And when they collide, the energy turns into heat and light and matter—
> GEORGE: Producing the world as we know it!
> . . .
> RORY. That would make string theory—
> GEORGE. A manifestation of—
> ALL. M-theory.
> RORY. (dreamily) M-theory.
> JUNE. The Mother of All Theories. (249–51)

Once again, the disarrangement of a domestic space gains substance through its association with an ostensibly novel cosmic order, revealed by string theory as scientific insight. The exchange represents a tentative closure, a suspension of animosity between the two male rivals, in love and in the work of what has been presented as their respective competing theories (that echo the accounts from popularizations). To "accommodate gravity," string theory expands from ten to eleven dimensions to become a "manifestation" of M-theory. M-theory, all three incant, becomes comprehensive and cohesive as the "Mother of All Theories." The expansiveness of "eleven dimensions" may accommodate a novel acceptance of a previously untenable, irresolvable interpersonal conflict. In the double significance, June, as gravity, may "accommodate" the validity of the perspectives of both rivals/spouses. The juxtaposition of act 1, where George is husband and Rory the illicit lover with act 2, where roles and perspectives are reversed, suggests a more tolerant world that endorses an imaginative empathy. That empathy naturally leads to a heightened capacity to forgive personal transgressions, in this case, marital infidelity. The ekpyrotic model imaginary, as a cosmic order, validates the rearrangement of a domestic order; the love triangle, in its exposition, becomes stable even as it subverts the domestic order that preceded it, with its anxious charade of marital fidelity.[33]

This interpersonal reconciliation coincides with or even results from the frisson of scientific collaboration. As such, the denouement of *Strings* also suggests a romanticization of scientific collaboration. As Shepherd-Barr observes of many contemporary science plays, in its economical use of props and scenery, *Strings* foregrounds both the text and the actors' bodies. This "de-emphasis on spectacle" serves a dual purpose in keeping with the reading I have just offered. Firstly, that the play is set in a train compartment draws attention to the intimate and the interpersonal; it reinforces the impression that what is at stake in the play is how three individuals cope with questions of their relation to each other—juxtaposed to an imaginary that concerns itself with their place in the broadest possible context, that of the universe itself. A domestic- and macro-scale space are closely coupled. Each character's ghostly familiar, the heroic scientist with whom he or she most closely identifies, appears at pivotal moments in the plot to mediate that connection between the interpersonal and the cosmic. Furthermore, a minimalist set, in some respects, gestures toward what is conventionally recognized as a pedagogic space—the classroom or lecture hall. In so doing, *Strings* attempts to elide lessons concerning string theory and "life's lessons," an enhanced sensitivity to the consequences of our choices in romantic relationships and the possibilities of empathy. Such empathy would promise interpersonal reconciliation through mutual recognition, acceptance, and the eventual alignment of creative energies toward a higher purpose. In the case of *Strings*, this ideal is expressed at the end of act 2 in the collaboration that takes place between the estranged threesome in the resolution of the three "theories of everything," the reconciliation of M-theory (George) and string theory (Rory) with gravitation (June) through the ekpyrotic model.

Nevertheless, in *Strings*, an ambivalent deference to the pedagogic space of string theory popularizations—where a concern with faithfully reproducing the authority of its imaginary is at times compromised by its own contradictions—works to undermine the openness of its treatment of a domestic space. Rather than acknowledging the somewhat incoherent heterogeneity of the popularization imaginary, the play calls upon that authority to substantiate its domestic rearrangement through a reimagined cosmic order. That juxtaposition is, in this respect, completely gratuitous.

Hillman's "String Theory Sutra"

Thus far this chapter has covered a sample of texts that, with the partial exception of *Strings*, suggest a relative failure on the part of imaginative writing to engage with string theory ideas in a rigorous way. Part of this failure is the inability these texts display to acknowledge and complicate the highly conventionalized epistemological conceits that popularizations employ to legitimize and substantiate strings and branes. They tend to simply reproduce them in an act of deference to the authority of the popularization. While I have encountered numerous poems that engage with string theory as a scientific imaginary, I want to focus in this last section on one poem in particular. This poem represents a sustained engagement with string theory in a way that not only satisfies all three of Shepherd-Barr's criteria, but also exceeds the functions of borrowing posed by Le Doeuff in that it does not seek to validate its own "enterprise" through patronage to string theory as authoritative scientific knowledge, nor does it reiterate the romance of power endemic to popularizations and many of the imaginative treatments of string theory considered in this chapter.

The poem is "String Theory Sutra" by Brenda Hillman.[34] By calling attention to the fundamental heterogeneity of a string theory imaginary, Hillman's poem successfully counters the tendency to reinforce the conventionalized epistemological conceits that find their most overt expression in popularizations. The poem achieves this by engaging a lyrical narrative "I"[35] with the basic image of the string, and through that engagement, frenetically exhausting the image's associative possibilities by a process of textual "stitching" together of what the narrator calls "inverted fragments" (168). The imaginative and formal coherence of the poem insistently and knowingly works against its own aspirations to an authoritative totality.

The poem's title itself suggests this flattening of a hierarchy into an associative multivalent topography, a stitching together of equally weighted images. *Sutra* is a Sanskrit word that means *thread* or *string*; it is the noun form of the verb *siv* which means "to sew." The literal sense of *sutra* as "that which is sewn together" was abstracted in antiquity eventually to connote "rule." The *OED* informs that in ancient Sanskrit literature a *sutra* was "a short mnemonic rule in grammar, law, or philosophy, requiring expansion by means of a commentary." Hillman suggests that the "inverted fragments" mentioned above are "like Bay Area poetry" (168). These "inverted fragments" refer dually to string theory as an imaginary and, concomitantly, the poem itself, which, the implication goes, is also like Bay Area poetry. Through this invocation, Hillman draws a filial connection between the poem and a Beat poetic tradition associated with San Francisco and Berkeley, California. Specifically, through its effusive, stream-of-consciousness form, "String Theory Sutra" imitates and comments upon the poem "Sunflower Sutra" by Allen Ginsberg, which also features verses that take the shape of inverted fragments:

> Unholy battered old thing you were, my sunflower O
> my soul, I loved you then!
> The grime was no man's grime but death and human
> locomotives,
> all that dress of dust, that veil of darkened railroad
> skin, that smog of cheek, that eyelid of black
> mis'ry, that sooty hand or phallus or protuberance
> of artificial worse-than-dirt—industrial—
> modern—all that civilization spotting your
> crazy golden crown—
>
> (36)

Here the image of the sunflower inverts and exhaustively muddles its easy association with a "holy," natural world set up in opposition to the "sooty hand" of industrial civilization.[36] Like "Sunflower Sutra," Hillman's poem complicates the all-inclusive authoritative pronouncement—the transmission of a certain cosmic knowledge—that a contemporary New Age culture fetishizes through the *sutra* as an emblem of an exotic Eastern epistemology. The somewhat ironic use of the term *sutra* effects its own alternative evocation of an authority that thwarts what New Age culture would identify as an Eastern mystical authority, an authority that ostensibly challenges reductive Western physics.[37]

Like Ginsberg's poem, "String Theory Sutra" is an *anti*-sutra, contradicting, as it does, the idealized clarity and precision of traditional Sanskrit sutras. The poem, in its excess, incorporates both the pithy maxim, in keeping with the spirit of a *sutra*, and a more long-winded array of images drawn from a wide variety of sources of received knowledge. String theory images and epistemological conceits are stitched together in the text through an associative peregrination with other received scientific knowledge. This amalgam is combined with an exegesis that links those ideas to the concerns of the lyrical "I" and, in effect, serves to explore the limits of the poem's and string theory's pedagogic authority. As such, the imaginary of "String Theory Sutra" is not simply a haphazard heaping of image upon image, but rather, a conscientious instantiation and figurative stitching of associative images. The imaginary of the poem becomes multidimensional, polyglossic, without being necessarily coherent or totalizing. It achieves a certain reconciliation of the radically heterogeneous with the ultimately unknowable.

String theory technical exposition employs the image of the string because it claims it to be an apt double for mathematical expression. The image finds its place in a complex, heterogeneous imaginary that allows theorists to imagine their own proxies interacting with objects in an abstract theoretical space through a specific proceduralism. This allows for an encounter with the alien, imaginatively structured in multiple, juxtaposed scales, that gives meaning to the endeavor in a way not possible with the purely conceptual. Crucially, the absolute valorization of concept over image in string theory technical exposition enables this imaginative work to escape a concerted scrutiny. Such scrutiny is simply irrelevant. "String Theory Sutra," on the other hand, works to unmask a lack of acknowledgement of the importance of the imaginary.

The poem does this by approaching the image of the string from the opposite tack as technical exposition. The poem begins with a declarative concerning the "personal in poetry":

> There are so many types of
> "personal" in poetry. The "I" is
> a needle some find useful, though
> the thread, of course, is shadow.
> In writing of experience or beauty,
> a cloth emerges as if made
> from a twin existence.
>
> (168)

Hillman draws attention to the multiplicity of the "personal" in the narrative voice. The "I" is synonymous with the "personal," which she states is a "needle some find useful." In the analogy, it is the "I" that is the tool that draws the figurative "thread" of "experience or beauty" into stitches, creating a "cloth" that coheres. It is significant that Hillman qualifies the personal "I" as only "useful" to "some." Obviously, all English speakers use the "I" as a matter of convention. This separating of a "some" from the all of those who use the "I" serves in part as a comment on the premium that a scientific discourse such as string theory puts on an objectivity that is categorically free of the personal. Hillman suggests that a poem such as "String Theory Sutra" does not claim for itself such objectivity. It embraces the multiplicity of the "I" in its performativity, while also acknowledging that the "writing" that results from its work cannot claim for itself the substantiality of an objective discourse—the "thread . . . is shadow." Nevertheless, a "cloth" does indeed emerge from this effort of the "I" to stitch "experience or beauty" with such a "thread." The poem then proceeds to explore the texture of that cloth.

The cloth, Hillman suggests, is "as if made from twin existence." The "as if" is key. The "twin existence" to which this utterance refers is ambiguous. It could refer to the "I" and the "thread" with which the "I" works. Or it could refer to "experience" or "beauty," which are distinct yet mutually informing. The implication is that while the work of the "I" in this poem would seem to generate a "twin existence," two entities that, like twins, are identical, or at least closely connected, that duality is, in fact, not stable. The poem undermines the certainty of its own declaratives; it repeatedly calls attention to the inherent ambiguity of its imaginary, without letting that imaginary deteriorate into an arbitrary assemblage of images. It announces that it will treat an imagined world that originates from an avowedly personal perspective and then extends out to engage with the associative possibilities of the image of string in a plethora of contextual valences. In so doing, the poem attempts to come to terms with the complexity of that world. Accordingly, the relationship between the "I" and string theory will not render itself readily coherent in accordance with the stock epistemological conceits of popularizations. If the cosmic order is a puzzle to be solved, string theory, as it is imagined here, does not promise a definitive answer. Furthermore, if "experience" and "beauty" are in a relationship of complementary opposition, analogous perhaps to the categorically distinct concept and image of scientific realism, then neither one nor the other constitutes an essence over an appearance, a "deeper truth." The "cloth" of reality arises from an interwoven combination of both.

In the sixteenth line of the poem, Hillman first mentions string theory itself:

> String theory posits
> no events when it isn't a

> metaphor; donut twists in matter—10
> to the minus 33 cm—its
> inverted fragments like Bay Area poetry;
>
> (168)

"String theory posits" imitates the expository language of science. But what completes this statement, after the considerable gap on the page, is a denial of causal interaction—"no events." The implication is that only when string theory is a metaphor does it posit events. In this epistemological imaginary, events emerge from a metaphorical displacement, where one can only recognize them through imaginative engagement.

The next image, "donut twists in matter," reiterates this notion in a complementary register. It is highly likely that Hillman adapts this image of "donut twists" from the chapter in *The Elegant Universe* where Brian Greene presents a complex donut-like figure meant to illustrate a six-dimensional Calabi-Yau space.[38] The donut, of course, is a hand-scale quotidian object: the image "donut twists in matter" bears with it a sense of the malleability, the tactility of matter as a dough-like substance. This tactility creates yet another shift in the imaginative register—from the conceptual abstractions of events and metaphor to a tangible image.

In the relevant chapter of *The Elegant Universe*, Greene explains that these "donut twists" do not occur, strictly speaking, in matter, but rather, in the geometry of space-time. He writes: "extradimensional geometry determines fundamental physical attributes like particle masses and charges that we observe in the usual three large space dimensions of common experience" (206). In this imaginary, one-dimensional vibrating strings constitute matter. As these strings move through a multidimensional space-time, that movement generates the particular vibrational resonances that, on subatomic scales, instruments measure as the particle attributes of mass and charge (the epistemological conceit of allotropy). It is doubtful whether Hillman intentionally distorts the image by placing the "donut twists" in matter rather than space-time. In any case, this is beside the point. It is precisely these kinds of so-called distortions that mark such images within a literary text as serving a function other than patronage or the repetition of a certain desire. A distortion such as this, rather than being a mere misunderstanding, is more a decontextualization—necessary and appropriate to the poem—that allows the image to be redeployed to an alternative imaginary. The poem makes no claim to authority over string theory as science—nor does it, unlike "On the Brane" or *Strings*, effect a deference to the authority of the popularization. Rather, what the distortion signals is a certain flattening of the imaginative space where authoritative voices—whether personal, scientific, historical, or other—are on an equal footing. There is no subordinating one to the other; within the purview of the imaginary, there is a parity taken to its logical extreme. The personal "I," through its "many types" and through its roaming over and linking up of images of strings in multiple contexts, engenders a fully realized polyglossia. In the "stitching" of one image to another from an entirely distinct epistemological domain, the disparate "threads" of string theory knowledge—whether an image of matter or of space-time—become conflated into an imaginative register that perpetually refers back to its own fundamental heterogeneity. As such, "String Theory Sutra" exemplifies a textual practice whereby an imaginary reconciles itself to its own multiplicity.

To return to the quote, the hand-scale image of "donut twists in matter" is immediately displaced into another fragment from string theory popularizations, "10 to the minus 33 cm." What was imaginatively just within reach becomes suddenly unimaginably far off. And yet, significantly, this factual "fragment" is partially unfurled in the text. On the one hand, Hillman uses Arabic numerals and the abbreviation "cm" for centimeters. On the other hand, the superscripted minus sign, conventional at least in popularizations, is written out as "to the minus." This may be a concession to the relative unfamiliarity of base ten logarithms to a readership who probably last encountered similar mathematical expressions in secondary school. Nevertheless, it provides a textual example of a kind of figurative kneading and flattening out, a curling and uncurling that can take place within such an imaginary, both in terms of form and of content. Hillman acknowledges this when she observes that "unexpected folds are part of form" (168). Form and content become mutually referential, and together, self-substantiating—an instantiated unity that is "as if made from a twin existence."

The line continues with "its . . . inverted fragments like Bay Area poetry." The possessive "its" is ambiguous. It may refer back to "string theory," but one could also make a case for it referring to "matter." The effect of this ambidextrous possessive is to reach back to link syntactically the one image to the other, to generate an ambient space where the metadiscursive idea of string theory, and the phenomenon it claims to substantiate, strings-as-matter, harmonize. String theory, as an abstract theoretical space, and matter come to consist of "inverted fragments." Phrases are arranged to approximate the warp and weft of fabric on the space of the page. As such, the poem itself is an aggregation of syntactically inverted fragments that formally reinforces the inversions, the intertwining of string theory images and their associative supplements, a weaving together. There is a certain capaciousness to "String Theory Sutra" due in part to the surplus of interstices between the images of string and the cascade of string-related associations that the poem generates.

Some of these associations represent an engagement with the pedagogic space of string theory popularizations, but also, importantly, with other sources of received knowledge. The personal "I" of the poem ranges over images derived from, among others, historical accounts of the textile industry: "decades after the spinning / wheel gathered stray fibers in a . . . whir of spindles before the swath / of the industrial revolution"; American military adventures abroad: "Nylon parachutes for / World War II. We shall not flag / nor fail, wrote Churchill"; labor movements: "After the workers' lockout 1922, / owners cut back sweatshop hours to . . . 44 per week"; patriotic rituals: "It's July / 4: air is full of mistaken . . . stars & the wiggly half-zeroes stripes / make when folded into fabric meant / never to touch ground ever again—"; and medieval mythologies: "Women, making weavings of / unicorns in castles, hummed as they sewed / spiral horns with thread so real . . . it floated." Their juxtaposition suggests a decentered worldview, a totality that calls into question its own totality. It is organized around the common thread of the image of the string, but also around a yearning for cohesion that motivates the personal "I" of the poem as a provisional center of affect amidst the shifting associations that link multiple scales, times, and contexts.

Alluding to a medieval imaginative tableau, the narrator declares, "O knight, . . . tie our scarf on your neck. . . . Over & inner & . . . code. The unicorn, *c'est moi.* The / rips by which strings are / tethered to their opposites like concepts . . . of an art which each example / will undo" (169, emphasis in original). The image of the knight, directly addressed, taking a scarf as a token of fealty suggests the romantic mode of patronage; the knight quests and battles as "our" champion. The knot in the scarf goes "over" and is "inner" both literally in the topography of the knot itself, and also abstractly, in the knot-like shape of reality. Coherence is perpetually undone by paradox, a weaving in and out of sense and nonsense. In this fabric, there is both concrete relationship and symbolic significance, both the material knot in the scarf and the knotting of concrete and abstract through "code." As in the passage, fragments from the received knowledge of history and string theory as science are radically juxtaposed.

The phrase, "The unicorn, *c'est moi,*" of course, refers to the famous, if apocryphal, assertion of Louis XIV, "*L'État c'est Moi,*" "I am the State," a succinct claim of absolute monarchy. In the altered expression, the presence of the "unicorn" calls attention to "*L'État*" in its absence and highlights the slippage of denotation in the substitution: nation state, psychological state, fantastic being. In effect, "I am the unicorn" calls attention to the "I" itself as an equally unstable, even mythological, imaginative amalgam. The implication is that, in this "tethering" of "opposites" that both conflates and disrupts ostensibly objective phenomena such as "strings" with mental abstractions such as "concepts of an art," categorical distinctions are undermined constantly. The fabric is both fluid and solid, its state perpetually shifting. In this context, unlike Louis XIV's decree, declarations of authority, while starting from the personal "I," must contend with this displacement, this macaronic complexity. To return to the image of the knight, whatever quest or conquest this knight undertakes simply cannot avail itself of the categorical purity of that epistemological conceit so common to popularizations—string theory as the domestication of a wilderness-like space. The poem's imaginary conscientiously acknowledges the familiar in the strange and the wild in the domestic—that all the received knowledge of our lived, yet thoroughly acculturated experience must be reinvented, reorganized, re-realized.

Within this profusion, this loosely woven (associative) matrix of string-related imagery, a declaration such as what follows evokes once again, subtly insists upon, the binding force of the intimately felt and feeling—the fragile, fleeting, and flitting "personal"—that permeates this imagined world and links it to the theoretical: "the slippage between / string & theory makes air seem invented" (169). To divide string theory into "string & theory" suggests again the "twin existence" of the first lines of the poem. Here "string" pertains to its sense as a human-scale, familiar image—the imaginatively basic, the graspable, workable string. "Theory" suggests the ostensibly intimidating intractability of string theory's conceptual content, which popularizations such as *Hyperspace* and *The Elegant Universe* have made accessible only enough to inspire a deferential gasp of astonishment, a certain wonderment. The "slippage" between the familiar and the technical, marked as intimidatingly alien, makes the very "air seem invented." Air is an apt image to capture this imagined "slippage" between the abstract and the concrete, the remote and the near-at-hand—connoting as it does, a substance both intimately and perpetually felt by the body and yet, conversely, a substance that by its nature is an absence, an invisible background or enveloping space through which our bodies move. In this imaginary, air connotes insubstantiality, and abstraction. Air indeed seems invented in this abstracted sense, this senselessness, much as the concept of multidimensionality does in string theory, an expansiveness that exceeds the comfort of the graspable, of common-sense.

In its insistent inversion of commonsense notions both of the personal and of received knowledge, the conventional ways that one imagines oneself in relationship with others and with the world as a whole, "String Theory Sutra" confounds any easy epistemological imaginary. In so doing, the poem implicitly calls into question the authority of string theory popularizations in the ready access they promise to string theory. Hillman ends the poem thus:

> Einstein called mystery
> of existence "the fundamental emotion."
> You were unraveled in childhood till
> you were everything. By *everything* I mean
> *everything.* The unicorn puts its head
> on your lap; from there it
> sees the blurry edge. How am
> I so unreal & yet my
> thread is real it asks sleepily ~
>
> (170, emphasis in original)

Einstein is the quintessential scientist-sage, the hero whose insights into the cosmic earn him a certain authority in popular philosophy—over the Big Questions in life. That "Einstein called mystery of existence 'the fundamental emotion'" speaks to the conventional conceit of the cosmos as a mystery to be solved. In string theory popularizations, this conceit finds its emotional grounding in the expectation that physics ought to inspire awe, a primordial openness to the "mystery" that is the universe, which should give way, in turn, to a longing fully to know it—that physics is, in one imaginative frame, the power to "solve," "decipher," or "harness" nature, akin to Kaku and Greene's wish. This is both an explanation and justification for scientific activity. Concomitantly, within the imaginary of the poem, the efforts of the "I" to come to terms with this so-called mystery become a stitching together of a cloth that emerges as its own form of knowing, a knowing that incorporates a matrix of the personal in addition to an array of observations derived from various forms of received knowledge, the most prominent of which being string theory.

But here Hillman claims an authority that allows her to declare knowledge of a "you" that may or may not be the reader. The link between this "you" and the reader may remain indeterminate inasmuch as distinctions between the "many types" of the personal no longer hold sway. The knowledge that the poem claims for itself engenders a radical destabilization of such categories. The "you [is] unraveled" to become "everything." Such an unraveling inverts the image of the stitching together of cloth. It is not the "you" that sees the "blurry edge" of the known world, but a "unicorn" with its "head on your lap." The "I" that then speaks is not the "I" of the implied author nor the "you" of the implied reader, but the "I" of the mythological unicorn. The poem ends with a question, uttered by the unicorn, that suggests that the "thread" is what is "real," rather than the imagined "I."

To those impatient for some form of coherent certainty in the poem, this insistent inversion of perspective and of frame may seem like superfluous obfuscation, playing as it does on the conventional conceit that life is a dream. But the imaginative inversions that constitute this poem are not gratuitous. What makes "String Theory Sutra" an achievement is, among other things, the way in which it refuses to defer to the authority of popularizations by complicating their commonsense epistemological conceits. The poem calls attention to the latent ambiguities of any neat sorting of concept from image, of objective from personal, of abstract from concrete. Furthermore, it challenges the recurrent *mélange* of string and brane images contextualized as essence/appearance, a depth below a surface, the goal of a quest, and the exploration and domestication of a wilderness-like space. In so doing, unlike the other texts examined in this chapter, it does not attempt surreptitiously to substantiate one bit of received knowledge in terms of another, to borrow an authorized cosmic order to legitimize a domestic space. "String Theory Sutra" demonstrates an acute awareness of the radical heterogeneity of string theory as a scientific imaginary, of its lack of cohesion, especially in its rapprochement with the interpersonal. It complicates in a productive way the ongoing dialogue in Anglo-American culture concerning the relationship between the cosmological and the social, the "out there" and the "amongst us."

Rereading, Binding

Inasmuch as a scientific imaginary serves as a cosmic order in the sense that I have defined it, such an imaginary says more about the "we" of the community and how that "we" hopes to define itself, to regulate its own social practice, than it does about the universe that it represents. What is highly significant about the adaptation of string theory as a scientific imaginary in these literary texts is not that such an adaptation distorts, but rather that it affords a means of imagining ourselves within and of the cosmos. This adaptive practice implicitly demonstrates that a certain cosmic order may influence, if not inform, a domestic space, and, quite possibly, vice versa. It further suggests that similar investigations into the transit of imaginaries from technical to popular discourses—for example, heliocentrism, evolution, or DNA—will afford a means of better appreciating the ways in which science overlaps with what I have suggested are religion-like responses to scientific knowledge, where I mean something more in keeping with the etymological sense of the word. According to the *OED*, Cicero equated the Latin *religion-em* with *relegere*, meaning "to read over again," where later authors root the word in *religare* meaning, "to bind together." In the context of string theory as a scientific imaginary, one can interpret its religion-like responses as both a rereading and a binding. Inasmuch as science, as a discursive practice, resembles religion as a rereading of images, as the circulation and collective adoption of a particular epistemological imaginary, it too serves to bind a community together through a given cosmic order duly embraced as both total and coherent.

In *Conversations on Science, Culture, and Time*, Michel Serres writes:

> The conception, the construction, the production of rapports, of relations, of transports—communication in general—evolve so fast that they continually construct a new world, in real time. We still live in a century or a universe of concepts, beings, objects, archaic statues, or even operators, while we continually produce an environment of fluctuating interferences, which in turn produce us. (114)

We have seen that, within these various discourses, the core images of the string and brane are relatively fixed and static. What makes them complex and dynamic is the ways that these various discourses link the basic images up through an array of epistemological conceits to a broader cosmic order. Based on the account given in this chapter, with few exceptions, contemporary Anglo-American literary treatments disappoint in their engagement with string theory. I blame this lack of sophistication in large part on the relative illiteracy of Anglo-American lay culture when it comes to understanding the limits and possibilities of a scientific discourse such as string theory. We are poor epistemologists when it comes to string theory and as a consequence, we tend to recycle unreflectively the archaic and simplistic epistemological conceits that circumscribe our conception of string theory insomuch as it constitutes a relationship with the cosmos as a whole. As Serres implies, an increased sensitivity to the crucial role that a scientific imaginary plays in our relationship with the world "out there" and concomitantly, with ourselves, surely would enhance our capacity to come to grips with the "fluctuating interferences" that, in turn, produce an "us."

CHAPTER 6

Strung Together

String Theory in Contemporary Globalized Culture

Yea but (quoth she) the perill of this place
I better wot then you, though now too late
To wish you backe returne with foule disgrace,
Yet wisedome warnes, whilest foot is in the gate,
To stay the steppe, ere forced to retrate.
This is the wandring wood, this *Errours den*,
A monster vile, whom God and man does hate:
Therefore I read beware. Fly fly (quoth then
The fearefull Dwarfe:) this is no place for liuing men.

(Spenser 44, emphasis in original)

As with many, I first encountered string theory when I picked up a copy of Brian Greene's popularization, *The Elegant Universe*, in 1999. Having at the time a passing familiarity with the counterintuitive idiosyncrasies of quantum theory, I was drawn to it by the prospect of discovering further strangeness lurking in the remotest recesses of the subatomic universe. Thus captivated by what I took to be the astonishing ideas of string theory itself, I paid little attention to the manner in which those ideas were expressed. In effect, I was not entirely cognizant of the book's patently romantic undertone. It was only in surveying a broad swath of string theory popularizations in the lead up to the writing of this book that I began to see a distinct pattern in their exposition—what I have called a heroic romance with an implicit, yet highly charged, fantasy of power.

Chapter 3 used Le Doeuff's distinction between concept and image to examine string theory technical exposition. In the triangulation that comprises string-theoretical praxis, mathematical arguments function as exact prompts for the experimental intervention in the physical cosmos. This intervention is analogous to touch, to bodily contact with the world. Yet by privileging mathematical argument as concept, as that which exclusively possesses truth-value, theorists are at liberty to displace affect into the exposition that surrounds and complements mathematical argument. That exposition is necessarily populated with images. By denying the conceptual status of this exposition, theorists are free to mix and match radically heterogeneous images in a multitude of scales and configurations in such a way that a maximum of possibility may be assured. Together these images form an imaginary that bears with it a cultural orientation—a system of meanings and values that serves to situate the theorists' imagined proxies within the abstract theoretical space, and thus structure interaction with that space.

Sharon Traweek's seminal ethnography of the professional particle physics community, *Beamtimes and Lifetimes*, confirms the predisposition among physicists to understand their working relationship with what they have come to formulate and thus understand as a cosmic order through the prism of romance. In technical exposition we see a muted version of the same imaginative regime. Theorists venture imaginatively to an abstract space, both remote and alien, and manipulate theoretical objects in a wide array of radically juxtaposed scales. Out of this work of construction, something, while still abstract, yet couched in autopoietic terms, emerges within and from the space—strings and branes. String theory technical exposition is structured in such a way that professional readers recognize, however tacitly, a theorist's presentation of images of strings and branes as an encounter with something not constructed. It is a discovery of nothing less than a novel cosmic order.

We also saw that certain conceits, dutifully repeated from text to text, reinforce this particular imaginary of encounter within string theory, and accordingly, reinforce this sense of a romance of power. These conceits vary in degree of abstraction. The most fundamental are the notions that theorists, and by extension, their audiences, can know the cosmos as an intelligible and coherent whole; that the cosmos consists of general kinds of things; and that we may know these things by the attributes that come to define them. Furthermore, fundamental to the epistemological imaginary that justifies and substantiates string theory is the conceit that string theory, as such, constitutes an allotropy. One substance—namely, the string itself—begets many forms, subatomic particles. Here we have an example of the subtle shifts in valence within a string theory imaginary that allow it to accommodate a wide heterogeneity of images and conceits. In one context, the string is a form comprised of a substance, energy that exhibits tension, and yet from another perspective, the string itself is the substance out of which subatomic particles such as the electron and the quark emerge.

One of the crucial ways that this abundance of conceptual valences is organized, tamed even, is by means of yet another epistemological conceit, so pedestrian as to be hardly noticed—that what is true is a depth below a surface. As we have seen, in the technical article on the heterotic string, one image and its context may be privileged over another in its truth-value by means of this conceit. When comparing the quark to the string, the string is simply the deeper truth, the one that must be found out. In the process of discovery of such truths, epistemological conceits begin to structure the interaction between theorists, as imaginatively projected human agents (proxies) in this abstract space, strings and branes (theoretical objects as real phenomena) that display their own agencies, and, by metonymic extension, the cosmos of which they form the essential, constitutive element. The discovery of said objects becomes the revelation of something hidden (as in, hidden dimensions), the solution to a mystery, the decipherment of a text, the construction of a model, a quest, a journey, an encounter with an other, and, in many instances, the domestication of a wilderness-like space.

It is in the context of this peculiar, but not so unfamiliar, string theory imaginary that Greene's decision to describe the universe in his book's title as "elegant" takes on an added significance. Greene is certainly not the first physicist to use this adjective. As Steven Weinberg explains, among mathematicians, traditionally, "an elegant proof or calculation is one that achieves a powerful result with a minimum of irrelevant complication" (134). What is arresting about Greene's expression is that it qualifies the universe itself as elegant, not the specific mathematical arguments from string theory that produce the particle spectrum that comprises the universe. Greene's expression conflates the universe with the mathematical formulas that would constitute, in the imaginary, its essence. That a mathematical proof can be considered "elegant" if it "achieves a powerful result with a minimum of irrelevant complication" is a matter of convention. But in light of the epistemological imaginary that informs both *The Elegant Universe* itself, and popularizations in general, "elegant" takes on a connotation more in keeping with the romance of power endemic to that text. That the universe is elegant, according to Greene, speaks to both the way that theorists imagine themselves interacting with the cosmos as an abstract theoretical space, and with the constitutive agents of the cosmos, strings and branes, idealized within that space as autonomous entities. As discussed, they are autonomous in that strings and branes are objects that exercise their own agency. Theorists must first imaginatively encounter strings and branes and then tentatively engage with them—in short, interact.

In a social context, the quality of elegance connotes refinement, grace, luxury. That the universe is elegant implies that it has, at the very least, been domesticated. It is known, and further, known to be intelligible, and ultimately hospitable to, or at least compatible with civilization (i.e., the anthropic principle). The cosmos is not, therefore, a bewilderingly hostile, remote, and alien space, but, to push the word's connotation to its logical, if not absurd, extreme, rather resembles a comfortable salon or, if one acknowledges the scale of it, perhaps a cathedral. What is so remarkable about this ostensibly innocuous qualifier is how far it retains the vestiges of a sensibility that seems drastically at odds with the radical novelty, the cutting-edge glamour of string theory as scientific knowledge. What confronts us in this qualification is a sensibility altogether archaic, more in keeping with the chivalric exploits of the Redcrosse Knight than, to recall Merz and Knorr Cetina, the "policies, ansätze, tricks and devices . . . mutually embedded in one another within a sequential interactional system involving disembodied objects, several physicists and competing teams" of string theorists (74). In the popularizers' efforts to make accessible the highly abstracted work of string theory, they find themselves relying more and more on a shared imaginary that resonates with a sensibility aptly described as medieval or, at the very least, early modern. However antiquated or naïve, this sensibility works to support and sustain both the activity itself, as well as the dissemination of its imaginative derivatives.

String Theory and Religion

Chapter 2 considered a passage from Sokal and Bricmont's *Fashionable Nonsense* where the authors question those in the humanities who have a propensity to make dubious use in their work of what they see as an "avalanche of ill-digested scientific (and pseudo-scientific) jargon" (155). At the time of its publication in 1998, the book represented a rejoinder in an ongoing debate concerning the proper adaptation of scientific knowledge, a debate fanned considerably by Sokal's much-publicized hoax article, "Transgressing the Boundaries: Towards a Transformative Hermeneutics of Quantum Gravity," in the prominent (at the time) sociology journal *Social Text*. With an authoritative tone, Sokal writes:

> Feminist and poststructuralist critiques have demystified the substantive content of mainstream Western scientific practice, revealing the ideology of domination concealed behind the façade of "objectivity." It has thus become increasingly apparent that physical "reality," no less than social "reality," is at bottom a social and linguistic construct; that scientific "knowledge," far from being objective, reflects and encodes the dominant ideologies and power relations of the culture that produced it; that the truth claims of science are inherently theory-laden and self-referential; and consequently, that the discourse of the scientific community, for all its undeniable value, cannot assert a privileged epistemological status with respect to counter-hegemonic narratives emanating from dissident or marginalized communities.

Sokal intended the article, and this passage especially, to be a parody of a philosophical position central to the strong program or constructivist school of sociology. Interestingly, among other ideas drawn from contemporary theoretical physics, Sokal uses string theory to justify his central assertion.

> In the 1980s a very different approach, known as string theory, became popular: here the fundamental constituents of matter are not point-like particles but rather tiny (Planck-scale) closed and open strings. In this theory, the space-time manifold does not exist as an objective physical reality; rather, space-time is a derived concept, an approximation valid only on large length scales (where "large" means "much larger than 10^{-33} centimeters"!).

The first sentence echoes the stock definition of string theory offered up by the numerous popular accounts. The crucial argumentative leap occurs in the second sentence, where Sokal slyly proclaims that string theory denies the status of the "space-time manifold"; that is, the cosmos itself, as an "objective physical reality." Instead, it is a "derived concept." This assertion alludes to the concept in theoretical physics of background independence but warps it such that the evocation of a stock summary of string theory's core idea serves to authorize the notion that reality is "a social and linguistic construct."[1] In an imaginative leap, a "derived concept" necessarily implies a "construct."

What is at stake in this passage is another, perhaps less obvious, example of an imaginative parataxis, where a cosmic order is radically juxtaposed to a domestic space in order to validate human-scaled social concerns. As with syntactical parataxis, the capacity for the reader to fill in the associative gaps between the juxtaposed ideas depends on a certain cultural literacy, a certain common-sense, that often is ideologically motivated. The parataxis in this passage contrasts the relatively proximal notion of "objective physical reality" with the remotely abstracted "space-time manifold," setting in opposition two expressions that exploit conventional expectations with regard to their comprehensibility. The "space-time manifold" is understood as not something that we may grasp physically; it is a "derived concept" in that a "manifold," as such, is an esoteric mathematical abstraction. This image is juxtaposed to a reality that is physical in the sense that it is something to which our bodies, through the conduit of our senses, have direct access. This commonsense notion of "objective physical reality" posits the belief that reality is something we directly experience. It is first and foremost the unmediated apprehension of the objects around us that are within reach. Common-sense tells us that objective physical reality is literally graspable.

Michio Kaku, in an explicit move consistent with Sokal's covert position, claims that the "purpose of science is to peel back the layer of appearance of objects to reveal their underlying nature" (*Hyperspace* vii). String theory as science supposedly overturns the commonsense apprehension of objective physical reality. What is close-at-hand in the domestic space of our immediate, intimate environs—our homes, offices, labs, and parks—is actually only appearance, under which lies a deeper reality, the reality of strings. And here we have Sokal's rhetorical sleight of hand: since space-time is an appearance derived from the reality of Planck-scale strings, it is akin to a "social and linguistic construct," the antipodes of that which is objective. An inverted copula—the statement "does not exist as"—effectively makes reality subjective, confirming what Sokal dubs facetiously the "post-structuralist counter-hegemonic narrative."

One irony of Sokal's bold revelation is that, with respect to the close reading of string theory popularizations and, to a lesser extent, technical exposition, when he insists that "scientific 'knowledge,' far from being objective, reflects and encodes the dominant ideologies and power relations of the culture that produced it," this book has found that the assertion indeed rings true. The way that theorists imagine themselves in relationship with phenomena at both the Planck and cosmic scales, the scouting out and pinning down of strings and branes, betrays a certain preoccupation with domination. As an imaginary, string theory discourse asserts a privileged epistemological status that duly institutionalizes the theorists' way of imagining in the abstract a particular configuration of power relations, acculturated through and through, concerning their imagined proxies and the objective phenomena with which they wrestle. As soon as one departs from the "pure" abstraction of mathematical argument, a departure that is absolutely necessary if that knowledge holds any promise of being at all comprehensible in any social context, one is confronted with the tricky proposition, not of what the cosmos means, but what the cosmos means specifically to us, in a material, historically contingent culture. That binding of feelings and values to action, that radical concatenation of the abstract with the concrete, of the outlying with the proximal, of the peripheral with the graspable, is arbitrary insomuch as its legitimacy, its commonsense, comes about ultimately through the sanction of an authority.

It is easy to imagine Sokal chuckling when he discovered that his outrageous assertions were met with uncritical approval by the editors of *Social Text*. As critics have observed, while such assertions mimicked a trivial version of arguments made by the strong program, the essay's positive reception was assured, undoubtedly, because it arrived under the mark of authority to which they felt compelled, however hesitatingly, to defer—the authority of a professional physicist. Sokal intended his hoax article to embarrass the sloppy peer-review practices of constructivist sociologists, blinded as they were, from his vantage point, by ideological bias. Yet in effect, "Transgressing the Boundaries" actually served to further reinforce disciplinary boundaries through its clandestine use of imaginative parataxis.

As a form of what Felicity Mellor calls "routine boundary work," popularizations present one particular concatenation of a certain cosmic order with a human-scale space that emphasizes accessibility and domestication. Chapter 4 showed that, in large part, the representation of string theory in literary texts reproduces the romance of power characteristic of popularizations as a discourse, and its complementary epistemological imaginary. This epistemological imaginary presupposes humanity in its entirety as a cohesive entity, as well as a notion of history as something both monolithic and progressing linearly. According to both Greene and Kaku, the power available to us, then, is twofold. On the one hand, according to Greene, it is the power to "ascertain and comprehend some of the most mysterious characteristics of the physical universe . . . to take in the view from the summit and gaze out on the vast and elegant universe with a perspective of infinite clarity" (117, 387). On the other hand, that power is material and economic; it is the potential cumulative capacity to "harness" the energy of the cosmos, the energy inherent in strings, to human ends. As we saw in the previous two chapters, this "vision" of string theory's potential is more often than not couched in quasi-religious terms.

In "How to Be Iconophilic in Art, Science, and Religion," Bruno Latour draws a distinction between the substantiation of objects that scientific practice effects and an alternative mediation that he calls "person making." Traditionally, Latour argues, the process of person making was the proper domain of religion, rather than science; he does caution readers.

> This has not always been the case however, and there used to be a time where the most common, public, and collective form of life was not information transfer but person making. I hesitate to use the word "religion" to describe this form of mediation, since religion has been turned . . . into something exactly *opposite:* a belief in the existence of a distant substance beyond the realm of experience to which we have access only through the intermediary of special vehicles—a definition that, funnily enough, is a good description of science production, but *not* of person making. (Jones and Galison 431, emphasis in original)

Further on in the article, Latour does indeed use religion as the term of choice to describe this act of person making, which I define here, in keeping with Serres, as a rereading that binds together "our rapport with things" and "our relations among ourselves" through an imaginative parataxis (*Conversations* 141). Latour's hesitation in invoking the highly charged term *religion* hinges on its participation in what he calls metaphysics: "a belief in the existence of a distant substance beyond the realm of experience to which we have access only through the intermediary of special vehicles." Yet it is just such a belief that is required from consumers of string theory popularizations, as well as most literary texts that adapt string theory ideas. The "special vehicle" that serves as the intermediary between the "distant substance" and the "realm of experience" is precisely the popularization itself as a text, predicated, as it is, on the tacitly accepted authority of its author, the string theorist.

At the beginning of *Hyperspace*, Michio Kaku tells the story, in overtly romantic terms, of the awakening he experienced as a boy to his calling as a theoretical physicist. The first incident, he writes, occurred at the Japanese Tea Garden in San Francisco where, as a child, he would crouch beside a pond, "mesmerized by the brilliantly colored carp swimming beneath the water lilies" (3).

> In these quiet moments, I felt free to let my imagination wander; I would ask myself silly questions that only a child might ask, such as how the carp in that pond would view the world around them. . . . Living their entire lives in the shallow pond, the carp would believe that their "universe" consisted of the murky water and the lilies. . . . I once imagined that there may be carp "scientists" living among the fish. They

> would, I thought, scoff at any fish who proposed that a parallel world could exist just above the lilies. To a carp "scientist," the only things that were real were what the fish could see or touch. The pond was everything. An unseen world beyond the pool made no scientific sense. (3–4)

For Kaku, the "unseen world" that he seeks to illuminate is the world of hyperspace, the extra-dimensional cosmos revealed by string theory. That unseen world is the deeper truth, above and beyond the "shallow pond" of "murky water." What is scientist-like about the carp living in the pond is their insistence that what is real is only that which the inhabitants of the pond "could see or touch." *Hyperspace* as a whole proceeds to make accessible to its readers that unseen world in a way that allows them to imagine themselves seeing and touching it. It is through this conduit of the pedagogical space that we too as a lay audience, beholden to the authority of Kaku, the string theorist, may have our own conversion experience, our own revelation of the wider, deeper, truer cosmos of string theory. Kaku wants his readers to imagine the medium of their understanding to be the "murky water" of a scientific imaginary; in particular, the four dimensional space-time of quantum theory and general relativity, which has preceded the one offered up by string theory.

Earlier I mentioned that inasmuch as one may imagine the universe in spatial terms as a cohesive whole, to describe it as "elegant" partly suggests architecture. It is not uncommon to hear the institution of science described as a cathedral, the Cathedral of Science. In conceiving a design for the Natural History Museum in London, Captain Robert Fowke is reputed to have coined the expression, shortly before he died in 1867. A century later, architect Robert R. Wilson claimed to have purposefully designed Fermilab's main building, in Batavia, Illinois, to recall a cathedral in Beauvais, France. Lawrence Krauss writes in *Seed Magazine* that "the late Austrian-American physicist Victor Weisskopf described the grand particle accelerators that began to take shape around the world in the 1950s and 60s as the 'Gothic cathedrals of the 20th century' ("Discovery"). Krauss goes on to observe that "the comparison was, and is, apt. The medieval cathedrals pushed the limits of available technology, involved the craftsmanship of literally thousands of skilled workers, and took generations (and sometimes centuries) to complete." Krauss's perspective here emphasizes the similarities in historically contingent social and material productions. But the article's subtitle also reveals a subtler parallel between twentieth- (and, by extension, twenty-first-) century high energy physics and one of the traditional functions of religion: "From the new particle accelerator at CERN may emerge answers to the most fundamental questions of the universe." If we permit ourselves the liberty of attenuating the analogy between the contemporary institution of high energy physics and that of the medieval Catholic Church, then the physicists, playing the role of priests, would "emerge" from the cathedral that is the Large Hadron Collider at CERN, bearing with them "answers to the most fundamental questions of the universe" for consumption by a lay audience, such as those who read *Seed Magazine*, playing the role of the cathedral parish. Within the analogy, physicists mediate, much as Catholic priests once did (and still do for many), between an elusive, potent, and mysterious other—what is "out there" as nature—and the culture, the "among us," that submits to the priests' authority as arbiters of this esoteric knowledge.

To indulge in one last etymology, it is interesting that the term *theory* bears traces of this emergence. The *OED* informs that, in addition to its original Greek sense as "a looking at, viewing, contemplation, or speculation," theory also signifies a "sight or spectacle" and, more specifically, it refers to "a body of *theors*" or priests "sent by a state to perform some religious rite or duty, a solemn legation." Bound up with the sense that a theory embodies a form of stable knowledge concerning an objective, external world is a commingled sense that theory also serves as a kind of religious spectacle or ritual, the enactment of which is meant to inspire both reverence and awe (and perhaps also obedience).[2]

The analogy, of course, has its limits. Taking their cue from scientific realism, string theorists would point out that, unlike theology, whose epistemological authority hinges entirely on texts, and on those anointed with the power to interpret those texts, high energy physics bases its authority on empiricism, on methodologies that insist on a rational consensus concerning the evidence experiments unambiguously yield. Yet it would not be excessive to agree with Serres when he suggests that "the rapprochement of scientific discovery and religious conversion is drastically and mutually illuminating" (*Conversations* 82).

Projecting a Cosmos: String Theory Production

In chapter 2, I discussed what Sharon Traweek calls the "Durkheim supposition," which posits that "a culture's cosmology—its ideas about space and time and its explanation for the world—is reflected in the domain of social actions" and vice versa (157). To borrow once again a term from computing, both Durkheim and Traweek see the relationship between social practice and cosmology as a *feedback loop*, an interaction where two epistemological domains mutually constitute each other. This book has focused on string theory as a scientific imaginary—on the ways in which string theorists imagine themselves into an abstract theoretical space that, in turn, comes to constitute a cosmic order, and how this socio-cognitive act of imagining gets adapted and, in some respects, repurposed by non-technical discourses. Accordingly, this approach has emphasized *imagined* social actions within those abstract spaces. Now, in concluding this book, I would like to explore, albeit briefly, the reverse: how a string theory imaginary may be shaped by social actions understood as *material*—on string theorists as embodied social agents operating within an organizational economy. This is a speculative exercise that seeks to reckon the extent to which string theory as an imaginary may, in fact, resemble the principle sites of its production.

In his "cultural phenomenology" of theoretical physics, Martin Kreiger compares quantum field theory, a contemporary of and conceptual cousin to string theory, to a factory.

> The workings of Nature are analogized to a factory with its division of labor. But here the laborers are of three sorts: walls, particles, and fields. Walls are in effect the possibility of shielding and separation; particles are the possibility of sources and localization; and fields allow for conservation laws and path dependence. (1)

Kreiger goes on to describe the physicist's problem as discerning the "the political economy" of this system.

> (1) to describe the precise modes or mechanisms by which objects are delineated and so separated from each other—the *walls*, shields, and surfaces; (2) the names or labels or properties through which objects have their own identity and are influential in the world—*particles;* and, (3) the provision and delineation of space with its own properties, so that in space's interaction with particles we have an account of Nature's workings—*fields*. (4, emphasis in original)

The division of labor within the "factory" corresponds to the physicists' dividing up of their conceptualizations of "Nature." For Kreiger, physicists "take hold" of the world by adopting the particular problems that the community of practice to which they belong deems worthwhile, and accordingly, by participating "in its practices, culture, and ideology, thus employing the conventional models and analogies" (74). Doing physics is a haptic practice: "Knowledge is handling . . . The Archimedean analogy not only describes the physicist's research work itself, but also the physicist's theoretical structures—*handles* being degrees of freedom, *probing* modeling our interaction with Nature, and *tools* often being physical models and mathematics as well as experimental equipment" (99, emphasis in original).

But I wonder: if the doing of theoretical physics is a haptic activity where the practices of the community correspond to the interacting components of the theory they collectively fashion, then perhaps it is more appropriate to analogize contemporary theories such as quantum field theory and string theory not to the outdated image of the industrial factory, but to a more contemporary economic organization. After all, while part of a broader culture organized around industrial production in factories, the quantum theorists of the early half of the twentieth century did not practice physics in factories, but rather in research universities. It would seem altogether too facile to simply analogize the social practice of theoretical physics to the predominant economic organization of a given era—or even that era's predecessor. One ought to take into account the actual institutions in which the physics is produced.

To better understand how American research universities, in particular, have come to have such influence over the shape and direction of string theory, let us consider from a broader historical perspective the growth of theoretical physics as a whole. Both Greene and Smolin estimate the string theory community's population to currently number around 1,000 active members, the vast majority of whom are affiliated with research universities. It is research universities, then, as the principle hosting institutions for the practice of theoretical physics, that could serve as an imaginative template for string theory. Considering only the practice of theory, the string theory community's closest competitor in the quest for a quantum theory of gravity would be the *loop quantum gravity* community, with a population of about 100. String theory is, in large part, dominated by the American theoretical physics community, and thus, American research universities, who sponsor the theorists' careers.[3]

In the nineteenth century, physics as a discipline was dominated by Europe, especially Germany. The nineteenth-century German universities excelled at valorizing the ideal of research as a worthwhile pursuit for its own sake, epitomized by the concept of *Wissenschaft*.[4] Professorial chairs, guided by the policies of a central ministry of education, dominated those institutions and protected what historian Roger Geiger calls "autotelic" graduate research departments (255). On the other hand, due to their reluctance to grant excessive authority to the federal government, the Founding Fathers stymied President Washington's initiative to establish a national university that could impose universal standards on the nation's nascent university system. Hugh Davis Graham and Nancy Diamond note that, as a consequence, American universities grew in the nineteenth century to become "a loose, sprawling, largely unregulated system that was decentralized, pluralistic, competitive, and vast" (24). The modus operandi of these universities, mostly unregulated and varying dramatically in academic standards, was in large part to cater to the marketplace, rather than to pursue *Wissenschaft*. This marketplace centered around a demand for vocational training as well as basic humanities education that the secondary school systems largely failed to provide.

For the most part, this state of affairs persisted until World War II, when the contribution of science to the war effort brought scientists a tremendous amount of prestige, political influence, and ultimately, money.[5] The high energy physics community, the "atom smashers," sat at the apex of this pyramid of prestige, having made the most spectacular and arguably pivotal contribution to the war effort, in the form of the Manhattan Project's end product, the atom bomb. Political leaders worldwide came to instantly appreciate the importance of research, not only for national security, but also in order to compete successfully in the increasingly knowledge-based economies of the future. President Truman, on the advice of the science lobbyist Vannevar Bush, signed the National Science Foundation Act in 1950, which, in keeping with American-style anti-federalism, incorporated safeguards against excessive centralized influence. Though governed by a board appointed by the president, the board itself represented a wide array of interest groups, including consumers and businesses, as well as scientists (Graham and Diamond 29). Furthermore, the funds distributed by the NSF went not only to applied research and government-administered laboratories, but to pure research. Perhaps most significantly, a process of peer review decided how the funding was distributed. What had been a handicap the preceding century and a half with respect to the production of world-class basic research now became an asset—the American research university system's pluralistic, decentralized, fragmented, and market-oriented organization.

With the infusion of the cash subsidies for research from the NSF, further increased after 1957 by post-*Sputnik* competition with the Soviet Union, an influx of relatively high-quality students from the GI Bill, and the recruitment of numerous top scientists as refugees from the devastated European academic and economic infrastructures, American universities grew rapidly and enjoyed what is often called a "golden age" up through the 1960s (Graham and Diamond 34). Throughout this boom, American universities were able to maintain a balance between the seemingly conflicting interests of a mass market for higher education and world-class research, between populist inclusiveness and meritocratic competition. As Graham and Diamond point out, the American university system's dominance emerged as a direct consequence of "a large, nationalized academic market united by common organizational forms and professional standards; and consequently, competition between the campuses for students and faculty and sources of funding" (11).

Meanwhile, in this postwar academic boom where the fetishization of pure science by popular culture seemed to grow in direct proportion to institutional growth, high energy physicists maximized their privileged position within the scientific community as a whole. Traweek writes that:

> In the new mission-oriented labs of World War II, high energy physicists learned to administer large interdisciplinary teams of researchers, manage huge budgets, and speak the language of government agencies. . . . high energy physicists have maintained personal ties and influence in Washington. At the same time, their organizational skills and political acumen have not gone unnoticed in the universities: the expansion of the resources of physics departments is the envy of other disciplines, and many senior high energy physicists have become university deans, provosts, and presidents. (2)

Throughout the 1950s and 1960s, the "statesmen" of high energy physics were able to leverage their organizational and political expertise to win funding and protect their functional autonomy, to, in effect, reinvent the nineteenth-century German innovation—the autotelic graduate department—pumping out PhDs with increasing frequency, as well as basic research, both experimental and theoretical.

But in the 1970s, Graham and Diamond point to "shifting demographic, economic, and political conditions" as the source of a relative stagnation in American higher education (84). It was a period that has been called the "age of survival," the "stagnant decade," and the "steady-state era" for American higher education (84). This period of socio-economic stagnation at the major U.S. research universities coincided, in certain respects, with the mood of resignation within the high energy physics community. For high energy physics, the timing could not have been worse. As described in chapter 1, at the time, the standard model, along with its direct descendants the various quantum field theories—had been approaching, for practical purposes, full maturity. This made it much more difficult for high energy physicists to convince the interest groups who were responsible for distributing research funding, whether the NSF or within the respective universities themselves, that the projects they supported, the increasingly less dramatic "problems" they proposed to solve, justified the expense. This, in turn, put pressure on the theorists, concerned as they were about their career security in a hiring market where they witnessed a dwindling demand for college and university teachers, to conform more closely to the community's orthodox definition of legitimate research (Geiger 255–56).

With the recovery of the American economy in 1982, this climate of stagnation seemed to change: the Reagan administration pushed through an appropriations bill that more than doubled the amount of funding available to the NSF for dispensation (Graham and Diamond 117–19). But the Reagan administration's largesse came with a catch. Perhaps in accord with its explicit agenda to win the Cold War, the advisory board was instructed to favor applied over basic research. While the Reagan administration supported the proposed Superconducting Supercollider, that initiative ultimately failed in the early 1990s to make its case. Research on potential defense technology, even Reagan's fantastical pet project, "Star Wars," had clear precedent. In this atmosphere of the 1980s, where funding flowed freely primarily to applied research, it was the experimenters who found themselves in the more tenuous position politically. High energy physics experiments, while closer on the spectrum to applied rather than pure research, are nevertheless, by a significant order of magnitude, much more costly than the practice of theory. Recall that string theorists for the most part work alone or in small group collaborations. Unlike the experimentalists, who rely on and compete for "beamtime" at massively expensive accelerator/collider facilities, theorists conduct the bulk of their work with far less costly resources: chalk boards, whiteboards, notepads, and since the mid-1980s, relatively cheap and widely available computing and telecommunications technologies. In the mid-1980s, experimental high energy physics found itself in an excluded middle between applied research that directly served national interests, such as defense or economic competitive advantage, and pure research, which could keep the populous theoretical physics community, a tangible legacy of the golden age in basic research of the 1960s, busy—and at a relative bargain. As Steven Weinberg writes of the early articulations of a theory of everything in the face of daunting incompatibilities between the standard model and general relativity, "This of course did not stop some theorists from constructing very unnatural theories . . . in accordance with the oldest rule of progress in science, that it is better to be doing something than nothing" (126).

Since the 1990s, from an economic and institutional perspective, this gap between applied and basic research arguably has expanded. Graham and Diamond suggest that "the threat of a prolonged downward spiral in federal R&D funding and research support as expenditures on entitlement programs, debt service, and deficit reduction squeeze out discretionary spending by the mission agencies" aggravates an already tense national ambivalence over the evolution of American research universities (214). In spite of the American system's widely acknowledged world dominance in research acumen, attacks have increased by the press and advocate groups for a wide range of peccadilloes, including, among other things, "frivolous courses and research, grade inflation, student cheating, faculty sinecures, corruption . . . , scientific fraud, and bloated administrations" (216). With the emphasis in the American university system on the marketplace, the faculty feel more and more acutely a pressure to justify research direction and concomitant expenditures before an ever-widening constituency of stakeholders: from the incoming students and their parents, to the undergraduates that represent potential recruits to graduate programs, to alumni and other patrons of the university endowments, to administrative superiors, to a federal government that provides vital supplemental funding through the NSF and other consumer-oriented programs such as subsidized student loans, to state governments that regulate university accreditation and dole out their own funding, to the media, and lastly, to public opinion, taken as an amorphous, fragmented, and fickle whole.

In this economic climate of intensifying liberalization, fragmenting markets, and privatization, theoretical physicists have found it ever more difficult to attract sponsorship, whether from private businesses, that, understandably, prioritize research on marketable technologies, or from peer-review groups beyond their own that, due to the fragmentation of the practice of theoretical physics into increasingly specialized domains, find it difficult to maintain the necessary interdisciplinary expertise to make informed judgments. Björn Wittrock defines the university, in its ideal form, as "a true universe of all relevant domains of discourse which altogether reflect the sum total of human knowledge." But perhaps like Landscape theory, the current university system runs the risk of becoming a rapidly propagating constellation of braneworlds, isolated from each other within incommensurable domains of discourse and no longer capable of maintaining an "epistemic and normative universalism" (360).

In this institutional milieu of fragmentation and incommensurability, the "emotional power of cosmology," as Traweek puts it, takes on, with string theory, an uncanny metaphorical undertone (2). String theory attempts to recompile the proliferation and fragmentation of fundamental particles into one coherent formalism by the subtle conceptual shift of extending those point particles out into an extra dimension. The multitude of quanta become the vibrational resonances of one object—the string. But string theory has yet to achieve this unification as strings multiply—figuratively speaking—along an extra, extended imaginative dimension, into *p*-branes, *D*-branes, zero-branes, KK modes, etc. Concomitantly, the string theories themselves fragment and propagate into the various multidimensional models, into M-theory, F-theory, braneworld models, Landscape theory, etc. Ironically, this conceptual fragmentation within string theory would seem both to reflect and confirm the fragmentation of its production within the community of theorists itself.

Roger Penrose suggests, at least by implication, that string theory finds currency as a form of belonging: "If you follow the group, then at least you will have the companionship of others, and you can talk to them about the surrounding architecture and share the excitement of the quest for your common goal" (889). String theorists can take comfort in knowing that their work holds the promise of binding them to a community of practice, one that is both relatively stable in its norms and hierarchies, as well as locked into a certain epistemological momentum. Of course, describing community in terms of membership as its own end runs the risk of platitude, since one could make this observation about any community. What distinguishes the string theory community from many other communities of practice is the imaginary through which membership manifests. Experimenters are bound practically to their instruments—the accelerators, colliders, and detectors; conceptually, they are bound, to a lesser degree, to the data that their instruments produce and record, and lastly, experimenters are bound to the theoretical framework within which the data is organized. Belonging for experimenters rests in their daily embodied interactions with equipment and instruments; the spaces of their practice take on a localized, tactile specificity. String theorists, in contrast, find membership in an abstract theoretical space where material embodiment is systematically subordinated to abstraction.

On the other hand, the abstract theoretical space of string theory practice offers an explosion in its degrees of freedom, deleterious, in Penrose's opinion, to the theory's potential inevitability, but auspicious to whatever ludic proclivities a string theorist may have (897–98). So many degrees of freedom promise a tremendous potential for imaginative and practical busy-ness—epitomized in the sheer excess of problems begging for consideration. All that needs to be done, then, for political survival, is to hone an equilibrium between the theory's exclusivity—how its peripheries are closed, and its accessibility—how its peripheries are opened, and significantly, opened to which particular influencers; what Felicity Mellor calls, "routine boundary work." This balancing act in theoretical practice is akin to the American research university's mission in general to negotiate the competing pressures of meritocracy and the demands of a mass market. In the current institutional climate where an excluded middle between applied and pure research stigmatizes experimental high energy physics, string theory can protect its boundaries by making its epistemological and methodological demands more opaque to its immediate competition, thereby short-circuiting any technically targeted criticism, while enhancing the allure of the theory's imaginative underpinnings, opening the discipline to its commoditization in the marketplace and thus to a continued widening cultural currency—in short, baffle the meritocracy while delighting the masses. As we saw in chapter 4, one the main vehicles for this is the publication and promotion of string theory popularizations.

In this context, a string theory imaginary finds its greatest currency in a consumer culture that puts a premium on the trafficking of scientific ideas. Scientific ideas constitute an intellectual surplus subject to an economy of exchange where the cachet of a given scientific idea stems from its novelty, its weirdness, its sexiness, its arabesque opacity, its distinctiveness or interest-value relative to the banalities of the quotidian and pedestrian (yet which, often enough, circulates to the point of itself becoming a banality). A string theory imaginary finds its appropriate place in an information culture—a culture that marks scientific knowledge as the last frontier, where the unveiling of the hidden essence of nature becomes the virtual domestication of the alien on incredibly remote scales, and where the bandying about of scientific ideas in popular discourse becomes yet another vehicle for social displays of status. It is arguably a narcissistic culture, where, as Manuel Castells suggests in *The Information Age: The Rise of the Network Society*, "we are just entering a new stage in which Culture refers to Nature, having superseded Nature to the point that Nature is artificially revived ('preserved') as a cultural form" (477).

Castells articulates the fundamental transformation that has occurred in the transition from an industrial economy to an informational one. Unlike a predominantly industrial economy, where cheap inputs of energy and raw materials lead directly to the mass production and distribution of affordable consumer goods, in an informational economy, "the source of productivity lies in the technology of knowledge generation, information processing, and symbol communication" (17). Cheap inputs of energy give way to cheap inputs of information that are, in turn, invested back into the production cycle for the generation of yet more information. Castells defines "informationalism" as "a specific form of social organization in which information generation, processing, and transmission become the fundamental sources of productivity and power, because of new technology conditions emerging in the historical period" (21n33). He sees the decisive historical period for the emergence of the informational economy as the 1980s, when deregulation and liberalization allowed for the reorganization of the telecommunications industry; in particular, the divestiture of ATT in 1984. Castells argues that "the availability of new telecommunication networks and information systems prepared the ground for the global integration of financial markets and the segmented articulation of production and trade throughout the world" (52).

One could trace the roots of the telecommunications reorganization during the mid-1980s, along with the emergence of information systems as an integral part of economic production, back to the postwar economic boom. At the time, even though the American economy, the dominant global economy, primarily consisted of industrial production as defined above, federal policies designed to foster knowledge-production through basic and applied research at the growing universities, as well as techno-military supremacy over the Soviet Union, led ultimately to the establishment of ARPANet in 1969 at several major U.S. university campuses. ARPANet spread slowly over the next two decades until it exploded in the early 1990s into the Internet—that quintessential embodiment of the information economy that the developed world increasingly has come to depend upon in the twenty-first century as it becomes further integrated with almost every aspect of the global economy. Originally designed as a decentralized, nodal network for distributed computing, the first users of the Internet—scientists at those first few top-level research universities—quickly enhanced the Internet's operating system to accommodate internodal communication, most notably email, but, of course, later for newsgroups, Gopher, and eventually the World Wide Web.

Given this transformation from an industrial to an information economy, perhaps it is most appropriate to say that string theory resembles not the factory, nor the university, as a brick-and-mortar institution, but the network, and in particular, what Castells calls the "enterprise network." For Castells, the enterprise network is, for the first time in history, the "basic unit of economic organization," as opposed to the individual (i.e., the entrepreneur or the entrepreneurial family), or the collective—whether the capitalist class, the corporation, or the state (170, 198). He defines the enterprise network as "*that specific form of enterprise whose system of means is constituted by the intersection of segments of autonomous systems of goals*" (171, emphasis in original). Enterprise networks cut across cultural, disciplinary, institutional, and geographic boundaries to form an autonomous and cohesive entity. Importantly, they are autotelic in that they are directed toward self-defined goals.

The string theory professional community possesses these features. It functions nodally (practitioners working alone or in small groups) across institutional and national boundaries. It is autotelic and goal-oriented, or in the vocabulary of the community, oriented toward the solving of legitimate "problems." As Traweek observes, by embracing scientific realism and enforcing its universalist social codes and norms, it attempts to transcend culture and inhabit an ahistorical, "supranational and supracultural" space of practice (78).

Castells goes on to describe the selective porosity of enterprise networks.

> Cooperation and networking offer the only possibility to share costs, and risks, as well as to keep up with constantly renewed information. Yet networks also act as gatekeepers. Inside the networks, new possibilities are relentlessly created. Outside the networks, survival is increasingly difficult. (170)

This book has explored some of the dynamics governing the opening and closing of the string theory professional community's boundaries, both imaginative and organizational. Castells argues that, although enterprise networks are not absolutely pervasive in the informational economy—networks exclude as much as they incorporate—they do serve as its dominant process, its core organizing dynamic. Accordingly, the information society as a whole, taking its cue from its core enterprise networks, "is constructed around flows: flows of capital, flows of information, flows of technology, flows of organizational interaction, flows of images, sounds, and symbols" (411). Taken in aggregate, Castells calls this interlaced network structure a "space of flows," which he defines as "the material organization of time-sharing social practices that work through flows" (411). By organizing itself across and through the other networks that constitute the total system, the enterprise network reduces that system, in all its expansive heterogeneity, to flows of information. Such a system is best described,not in terms of the mutual attraction of rigid Newtonian bodies, nor the wave/particle ambiguity of quantum theory, nor the curved space-time of general relativity, but as a synthesis of them all. In an allotropic flattening reminiscent of string theory, where myriad subatomic particles reduce to the string, the disparate activities and interactions within the enterprise network reduce to information exchange. Furthermore, for Castells, within the enterprise network, force—the power that animates the system—is expressed in terms of the transformations of the topological contours of multidimensional spaces. In effect, enterprise networks constitute and are constituted by flows of information, just as matter and force are constituted by strings. To stretch the analogy to its limit, like string theory, the enterprise network projects a cosmos where the curving topologies of a finely granular multidimensional space privilege harmonic resonance over discrete objects and their colliding trajectories.

Castells does recognize, though, that this space of flows is not without its ambivalences.

> The dominant tendency is toward a horizon of networked, ahistorical space of flows, aiming at imposing its logic over scattered, segmented places, increasingly unrelated to each other, less and less able to share cultural codes. Unless cultural *and physical* bridges are deliberately built between these two forms of space, we may be heading toward life in parallel universes whose times cannot meet because they are warped into different dimensions of a social hyperspace. (428, emphasis in original)

Another interesting parallel between string theory as an imaginary and Castells's formulation of the space of flows is their implicit valorization of ahistoricity. For Castells, in the enterprise network, ahistorical, universal codes replace local, historically specific ones. One key goal for string theorists, as discussed in chapter 1, is to formulate the theory in such a way that space-time emerges from the formalism as a logical inevitability, not an assumed precondition. String theory strives to be ahistorical in a specifically technical sense, in that its mathematical formulation would supersede and subsume time itself. Such an ambition on the part of string theorists is entirely consistent with the universalist, ahistorical ethos that, as Traweek observes, they tend to embrace.

Tellingly, though, in his discussion of the space of flows, Castells evokes the concepts "parallel universes," "warped dimensions," and "hyperspace"—terms he surely appropriated from Michio Kaku's *Hyperspace*, a work Castells mentions earlier (376). Castells employs the image of a multidimensional "social" hyperspace to describe a materially grounded social phenomenon, the network enterprise, and its space of flows. On the one hand, such an analogy risks the kind of parataxis of disparate epistemological domains, of the cosmic and the domestic, so prevalent in literary adaptations of string theory. It threatens to call damning attention to the associative arbitrariness of this entire speculative exercise. A critic of these kinds of imaginative appropriations, such as Sokal, undoubtedly would complain that they poach technically precise, context-specific terms and arbitrarily inject them into irrelevant and misleading contexts. On the other hand, this evocation of string theory imagery on the part of Castells in the context of an explication of enterprise networks speaks to the tangled circularity of imaginaries in general. It suggests that string theory as social practice, both material and imaginative, generates string theory cosmology, which in turn generates social practice. It suggests that, in terms of the feedback loop suggested by Durkheim's supposition, it is entirely possible that string theorists, in their formulation of a multidimensional cosmos, adapted that image from a non-technical source—or that string theory mimics the enterprise network in which it is produced.

Another point: as we have seen, some critics decry string theory's overreliance on mathematical formalism at the expense of experimental evidence. In the past decade and a half, the Internet has become the preferred medium of communication for enterprise networks in general, and concomitantly, the string theory professional community. Merz and Knorr Cetina observe that theorists, when dislocated, communicate principally through email. Arxiv.org has grown quickly to be the dominant forum for the exchange of preprint technical articles. The Internet serves as the supporting material and virtual architecture of the abstract space of string theory. As such, it validates the virtuality of that abstract space. The informational space of flows of the community as a whole inform the multidimensional informational spaces of the networks of communication the theorists utilize in doing their work which, in turn, inform the multidimensional informational spaces of the theory itself.

The string theory professional community, in keeping with the dynamics of the informational economy as a whole, has been roaming, so to speak, within the theory's vast and highly specialized imaginative space. Throughout the past four decades, as string theory has metamorphosed from hadronic to bosonic to the myriad superstring versions to M-theory and its offspring, its imaginary has come to emulate and inform, in a process resembling a feedback loop, the imaginative conceits of the broader social practices that underpin and support it, the enterprise networks of research universities themselves, as well as the encompassing informational economy.

Throughout this book I have argued that string theory's epistemological coherence does not rest exclusively with the conceptual content that partially constitutes its technical discourse, but rather, must include an imaginary. An imaginary is not simply peripheral and supplemental, to be dismissed as part and parcel of the context of scientific discovery. For string theory, as a scientific discourse, both concept and image are mutually informing and interdependent. The centrality of an imaginary in the formation of string theory as scientific knowledge necessarily engenders an epistemology that accounts for physical embodiment, cognitive processes, and cultural practices that mediate human interaction with an "objective" world. Such an epistemology discredits the self-perpetuating myth of the disciplinary chasm between the "two cultures," in large part, by obviating the fraught distinction maintained by scientific realism between observation and construction, between the cosmic and the social.

Notes

Chapter 1

1. CERN is the *Organisation Européenne pour la Recherche Nucléaire* (European Organization for Nuclear Research). It is located on the border between France and Switzerland, just west of Geneva. It is home to the Large Hadron Collider.

2. This formula is called the *Euler beta function*, after the eighteenth-century Swiss mathematician Leonhard Euler. See Veneziano, "Construction of a Crossing-Symmetric, Regge-Behaved Amplitude for Linearly Rising Trajectories." Henceforth, endnotes will cite the full name of technical articles in order to provide context. For all other texts, endnotes will only cite details sufficient for readers to refer to the bibliography.

3. See respectively: Nambu, "Quark Model and the Factorization of the Veneziano Amplitude"; Susskind, "Dual Symmetric Theory of Hadrons I"; and Nielsen, "An Almost Physical Interpretation of the Integrand of the N-Point Veneziano Model."

4. This scale is known as the Planck scale, after the German physicist Max Planck. Of the Planck scale, John Schwarz writes: "one way that it is sometimes expressed is to say that the Planck scale is to the size of an atom as an atom is to the size of the solar system" (*Superstrings* 71).

5. John Brockman's promotion of a "third culture" provides a contemporary example of the supposed "moral" authority of scientists in general: "The third culture consists of those scientists and other thinkers in the empirical world who, through their work and expository writing, are taking the place of the traditional intellectual in rendering visible the deeper meanings of our lives, redefining who and what we are" (17).

6. In *Dreams of a Final Theory*, Steven Weinberg writes: "One common feature of everyone's idea of reductionism is a sense of hierarchy, that some truths are less fundamental than others to which they may be reduced, as chemistry may be reduced to physics" (51).

7. I originally used this term in a 2007 essay entitled "Imagining Braneworlds in String Theory Technical Discourse." Anneke Smelik also uses it in her edited collection, *The Scientific Imaginary in Visual Culture*, published in 2010. The book "explores the ways in which visual culture represents and remediates science" (9).

While Smelik contends that the term "'scientific imaginary' . . . indicates that science has profound effects upon the imagination, and conversely, of the imagination in and upon science," the emphasis is largely on the former.

8. Popularizations as a category also include semi-technical articles in magazines such as *Scientific American,* usually authored by journalists who specialize in writing about science, as well as articles by non-specialist journalists in mass media publications such as the *New York Times* or the *Guardian.* I will be focusing solely on monographs.

9. In some instances, mathematical formulae are incorporated into the text—what I call a semi-technical account; for example, Penrose's *The Road to Reality.* Others include equations in endnotes or an appendix. But even so, by extracting these mathematical expressions from their original argumentative context, do they not lose their precise significance? Do they become what Robert Laughlin calls "baubles"? "All of us have a powerful instinct to collect things that are 'interesting' even when they are useless" (133, 136).

10. In the manuscript for her play *String Fever,* Jacquelyn Reingold also acknowledges Greene in particular.

11. Tensor algebra is an algebra that describes the relations between arrays of quantitative information bound to geometric spaces: in the case of general relativity, tensors of mass-energy density and space-time curvature. Riemannian geometry, unlike its predecessor, Euclidian geometry, allows for the articulation of spatial curvature. It was developed by the nineteenth-century German mathematician Bernhard Riemann.

12. In *Hyperspace,* Michio Kaku defines a *field* as "a collection of numbers defined at every point in space that completely describes a force at that point" (25).

13. In *The Fabric of the Cosmos,* Brian Greene describes space-time as a kind of "loaf" in order to illustrate how space-time may be "sliced" in different ways—an analogy that helps to clarify certain temporal paradoxes that arise with special relativity (138–39).

14. Velocity measures the motion of a body along a direction in space; or, in the internationally standard terms of measurement, meters per second. Momentum represents the product of rest mass and velocity. Acceleration measures the rate of increase in velocity with respect to time, as in, for example, a massive body falling to earth, measured as a constant acceleration of approximately 9.8 meters per second squared.

15. Quantum theory as a whole is often divided into subcategories to distinguish advances or modifications. For example: quantum mechanics, quantum electrodynamics, quantum chromodynamics, the standard model, and the various other contemporary quantum field theories, which will be described in more detail later in this chapter.

16. Particles: little "parts" of the whole. The standard model is a quantum field theory that describes three of the four fundamental forces of nature: electromagnetism, the strong nuclear force, and the weak nuclear force. It comprises a "spectrum" of fundamental particles organized into two basic categories: fermions and bosons. Simply put, fermions are particles of matter and bosons are particles of force. Developed in the early seventies, the standard model has been validated ex-

haustively by a wide range of experiments. It has thus become the canonical theory among high energy physicists.

17. Note that the terms *flavor* and *color* are figurative. A subatomic particle is not literally red.

18. "Wavefunction" is conventionally written as one word. The square of the magnitude of the wavefunction can describe, for example, the chance that a particle has of being located at a given position or, conversely, of having a certain momentum.

19. In computer programming, the term *kluge* denotes a clever, ad hoc solution to a particularly extreme aporia. Kluges often approach problems tangentially by cobbling together a hodgepodge of provisional "quick fixes" into a tenuously persisting solution. String theorist Barton Zwiebach writes: "Quantum mechanics is a framework, more than a theory" (4). To be legitimate, other physical theories (including string theory) must be "quantized" (i.e., must be made to be consistent with the constraints of the quantum theoretical framework).

20. The standard model is organized into three symmetry groups, one having to do with the strong nuclear force, one with the weak nuclear force, and one with electromagnetism. Symmetry, in this context, means that the particles of a given group conform to a specific set of rules of transformation, akin to a rotation along an axis. Generally speaking, the more axes, the more particles in the group.

21. Here is one such constant in quantum theory: "The classic example of a coupling constant is the electromagnetic fine-structure constant α [approximately 1/137]. This dimensionless coupling constant controls the strength of the electromagnetic interactions" (Zwiebach 260).

22. Renormalization is a mathematical procedure whereby particle configurations are rescaled when momentum values get extremely large or, conversely, when distances get indefinitely small. The standard model is renormalizable, as is quantum electrodynamics. Quantum chromodynamics is also renormalizable, but that process alone is inadequate for making the model work. Most quantum field theories are not renormalizable, and thus, considered by most theorists to be ill-suited to describe physical reality. For a semi-technical explanation of renormalization, see Penrose 675–79; for a textbook exposition, see Zee 145–92.

23. The physicist Murray Gell-Mann devised the term *quark*. Reputedly, he was inspired by a phrase in James Joyce's *Finnegans Wake*, "Three quarks for Muster Mark" (383).

24. Bosons are the group of messenger particles of force, for example, the photon for the force of electromagnetism, the gluon for the strong nuclear force, and W and Z bosons for the weak nuclear force.

25. This was not the first well-known instance of a theorist positing the existence of extra space-time dimensions. In 1919, German mathematician Theodor Kaluza, in a paper he sent to Einstein, attempted to incorporate electromagnetism into general relativity by increasing the space-time dimensions from four to five (an idea first attempted by Gunner Nordström in 1914). In 1926, this idea was modified by Swedish mathematician Oskar Klein when he proposed that the fifth spatial dimension was undetectable because it was microscopically curled-up. In general, attempts to unify fundamental forces through extra dimensions are known

as Kaluza-Klein theories. Early versions of Kaluza-Klein theory were found to contradict the, at the time, state-of-the-art experiments in quantum mechanics. It is also worth noting that physicists frequently make use of extra-dimensional mathematical models to articulate the dynamics of systems of interrelated information beyond simply the positions and momenta of particles. Hamiltonians and Riemannian geometry, for example, can readily accommodate a multitude of dimensions. (In classical and quantum mechanics, a Hamiltonian is a mathematical operator that quantifies the total energy of a given system of particles.) For a textbook proof of the need for twenty-six dimensions, see Zwiebach 206–21.

26. Bosonic string theory not only calls for the existence of the tachyon, it is the theory's "ground state," its lowest energy or vacuum state. See Zwiebach 236–42.

27. Green and Schwarz, "Superstring Field Theory"; and, as a follow-up, Green and Schwarz, "Anomaly Cancellation in Supersymmetric D=10 Gauge Theory and Superstring Theory." See Greene, *Elegant* 138.

28. Generally speaking, fermions, which include protons, neutrons, and electrons, are the grouping of quantum particles that compose matter.

29. Developed initially in 1976 by theorists Daniel Z. Freedman, Peter van Nieuwenhuizen, and Sergio Ferrara, supergravity is a field theory that integrates supersymmetry with general relativity. Like string theory, some versions posit extra dimensions. But unlike string theory, supergravity posits point-particles rather than one-dimensional extended objects.

30. One of the proposed tasks for the Large Hadron Collider at CERN is to search for sparticles.

31. These theories vary generally by: whether or not they incorporate *chirality* (an asymmetry observed in the quantum particle world); whether they incorporate open and/or closed strings—loops or strings with loose ends, so to speak; and how they organize the particle symmetry groups. All of these theories, though, do call for ten space-time dimensions and for the compactification of the six extra spatial dimensions. For a textbook exposition, see Green, Schwarz, and Witten 291–352.

32. Witten, "String Theory Dynamics in Various Dimensions."

33. Perturbation is a method for finding an approximate solution to a mathematical problem that cannot be solved exactly. See, for example, Penrose 680, 922. Witten himself is less than forthcoming on exactly what the "M" stands for, providing a diversion for theorists as they speculate on its meaning. He offers three possibilities: magic, mystery, or membrane. Others have proposed: mother (as in "mother of all theories"), monstrous, matrix, an upside-down "w" (i.e., Witten), missing, and murky. Critics would argue that, ironically, M-theory itself still suffers from an excess of degrees of freedom, just like its name. See Woit, *Not Even Wrong* 155.

34. This has to do with how the rules of commutation are formulated with respect to the extra dimensions, or in other words, how a string rotates through space-time. Simply put, commutation is an algebraic relation that is transposable (e.g., $a \times b = b \times a$).

35. For a discussion of the origins of the brane as a theoretical object, see Miller 83–84.

36. Apparently, the "F" stands for "father," as a complement to M-theory's "mother." Vafa, "Evidence for F-theory."

37. From 1995 to February 2006, arXiv.org, the preprint clearing-house for many branches of physics, published over 33,000 preprint papers in the high energy physics/theory subsection, not including cross-references. The "hep-th" subsection devotes itself to "string/conformal/field theory" preprints. Not all the papers focus on string theory, but the majority do.

38. Maldacena, "The Large N Limit of Superconformal Field Theories and Supergravity."

39. Strominger and Vafa, "Microscopic Origin of the Bekenstein-Hawking Entropy."

40. Polchinski, "Dirichlet-Branes and Ramond-Ramond Charges." Instead of extending out into extra spatial dimensions, a "time-like membrane" extends into a dimension of time. Recall that, in keeping with the dictates of special relativity, within string theory, time is quantified as a space-like dimension. "D-branes" are named after the nineteenth-century German mathematician Johann Dirichlet, who pioneered work on boundary conditions for differential equations.

41. For a popular account of this "ekpyrotic" model (from the Greek, meaning "out of fire") of the Big Bang, see Steinhardt and Turok 121–66.

42. For an exposition of some of these concepts, see Greene, *Fabric* 415–94.

43. In *The Road to Reality*, Penrose writes, "String theory operates simply with a smooth 'classical' background spacetime, which is not even influenced directly by the presence of a string—since the basic unexcited string itself carries no energy, and so does not directly 'curve' the background spacetime" (893).

44. Greene writes: "Shrinking smaller than the Planck scale would be off limits not because you run into a fundamental grid, but because the concepts of space and time segue into notions for which 'shrinking smaller' is as meaningless as asking whether the number nine is happy" (*Fabric* 351).

45. For a sample of this kind of research, see Banks, Seiberg, and Shenker, "Branes from Matrices."

46. As mentioned in a previous endnote, commutation is a transposable algebraic relation: $a \times b = b \times a$. A comparable anticommutation would take the form: $a \times b = -b \times a$. In noncommutative geometry, the metrics of Euclidean and Riemannian geometry are replaced by matrices that do not commute. See Connes and Berberian 1–32.

47. Chapter 33 of Penrose's *Road to Reality* provides a semi-technical summary of twistor theory (958–1009). In short, twistor theory maps geometric objects in Minkowski space-time onto a particular complex four-dimensional space with special features. The following is Witten's earliest paper on the topic: "Perturbative Gauge Theory as a String Theory in Twistor Space."

48. The Large Hadron Collider is designed to operate at approximately 14 TeV or tera electron volts. The Planck scale energy is on the order 10^{19} GeV (giga electron volts). One tera electron volt equals 1,000 giga electron volts (i.e., a trillion, as opposed to a billion).

49. See Chown 42.

50. See Vilenkin xxi–xliv; and Kibble, "Cosmic Stings Reborn?"

51. See Hoyle, et al., "Submillimeter Test of the Gravitational Inverse-Square Law: A Search for 'Large' Extra Dimensions."

52. The following are some notable responses to string theory: Penrose, "Su-

persymmetry, Supra-dimensionality, and Strings" 869–933; Woit, "Is String Theory Even Wrong?" and *Not Even Wrong* 161–212; Krauss, *Hiding in the Mirror* 173–242; Smolin, *The Trouble with Physics;* interview with Sheldon Glashow in *The Elegant Universe: A Three-Hour Miniseries with Brian Greene;* Anderson, "God (or Not), Physics and, of Course, Love: Scientists Take a Leap."

53. In its defense, physicists such as Steven Weinberg anticipate that string theory, with its radical reconceptualization of fundamental particles, as well as space-time itself, may eventually suggest new ways to formulate experiments such that current technological limitations can be overcome. See Weinberg 212–19.

54. See, for example, Susskind, *Cosmic* 343–76.

55. In *The Cosmic Landscape,* Susskind writes: "The Landscape is *not* a real place . . . It's a mathematical construct, each of whose points represents a possible environment or, as a physicist would say, a possible *vacuum*" (90, emphasis in original).

56. For a discussion of this, see Krauss, *Hiding* 237–41.

57. Recall that perturbation is a mathematical procedure for finding an approximate solution to a problem that cannot be solved exactly. Certain problems in theoretical physics cannot be solved with a perturbative method. They require a precise solution, not an approximation.

58. Most versions of string theory assume a background space-time that must be "plugged into" the model as a starting assumption. String field theory attempts to derive space-time from calculations, obviating the need to insert space-time into the equations as an initial condition—as such, it is background-independent.

59. To reiterate: the experience of acceleration relates directly to—or, in other words, feels exactly like—the force of gravity.

60. Krauss paraphrasing Witten in *Hiding* 248.

61. See *Hiding* 191.

62. Ibid., 191, 202; and Penrose 890–91, 897–902, 926–29.

63. This "I know it when I see it" argument also speaks to the notion that, echoing the radical subjectivism of Feyerabend, science is what a culture collectively marks as science.

Chapter 2

1. This argument cuts both ways, as the notorious "Bogdanov affair" demonstrates. In 1999 and 2002 respectively, the Bogdanov brothers, identical twins, were awarded doctoral degrees from the University of Burgundy for theses that claimed to make original contributions to string theory cosmogony. In 2002, a controversy arose when physicists claimed that papers the Bogdanovs had published in reputable peer-reviewed journals, based on their research, were in fact illegitimate. Upon closer scrutiny, John Baez, in particular, found the papers to be "a mishmash of superficially plausible sentences containing the right buzzwords in approximately the right order. There is no logic or cohesion in what they write." The Bogdanov affair calls attention to problems that have arisen with the hyper-specialization within theoretical physics, which makes competent peer review all the more difficult. "Ill-digested scientific jargon"—fashionable nonsense—can contaminate even the scientific disciplines themselves.

2. Interestingly, contemporary satellite-based global positioning systems (GPS) rely on the application of special and, in some cases, general relativity for accuracy —a fact that has enjoyed some notice in the popular press.

3. See, for example, Penrose 493–524; and Greene, *Elegant* 97–102, 105–8.

4. Hosted by Cornell University, arXiv.org is a clearing house for pre-print technical articles by physicists.

5. As a self-avowed feminist philosopher, a crucial aspect of Le Doeuff's project is to contest the incoherences of Western philosophy as a male-dominated tradition. While my analysis of string theory technical discourse, as well as its popularizations, certainly suggests potentially feminist criticisms, a fully feminist reading of string theory as a cultural phenomenon is beyond the scope of this book. Furthermore, in keeping with Le Doeuff, my use of the term *imaginary* is distinct from its role in Lacanian psychoanalysis.

6. While principally concerned with philosophy, in a chapter entitled "Galileo or the Supreme Affinity of Time and Movement," Le Doeuff uses her notions of image and concept to investigate the epistemological and cultural consequences of Galileo's revolutionary theory that the speed of a freely falling body is proportional to the time elapsed (29–44).

7. See, for example, Bachelard, *Space* xvii–iii, and *Reverie* 176.

8. While such bold claims may not be categorically verifiable, Lakoff and Johnson base them on empirical evidence from cognitive psychology and neuroscience—on the convergence of evidence from a variety of experiments employing a variety of methods.

9. Generally speaking, analytic philosophy has championed the effort to equate mathematics with formal logic. An in-depth discussion of this undertaking and its nuances is beyond the scope of this book.

10. While corresponding to abstract mathematical expressions that pertain to scales impossibly remote from human apprehension, string theory imagery, even in its technical exposition, also bears with it a human-scale materiality.

11. In "Drawing Things Together," Bruno Latour also notes the importance of scale in scientific practice: "The *scale* of the inscription may be *modified* at will, without any change in their internal proportions. . . . Billions of galaxies are never bigger, when they are counted, than nanometer-sized chromosomes; international trade is never bigger than mesons; scale models of oil refineries end up having the same dimensions as plastic models of atoms. Confusion resumes outside a few square meters. This trivial change of scale seems innocuous enough, but it is the cause of most of the 'superiority' of scientists and engineers; no one else deals only with phenomena that can be dominated with the eyes and held by hands, no matter when and where they come from or what their original size" (Woolgar and Lynch 45, emphasis in original).

12. For a detailed discussion of the container as an image with a schema, see Lakoff and Johnson, *Philosophy* 27–36.

13. It is important to distinguish cosmology as an imaginary, and cosmology the science. Within physics, cosmology enjoys a special status as a distinct discipline. Cosmologists are concerned with understanding the universe on the largest of scales by means of theoretical models extrapolated from, among other methods, telescopic observation: the dynamics of solar systems, black holes, galaxies, and gal-

axy clusters. While they work from the assumption that the universe on the vastest of scales can be understood as an ordered whole, they would be quick to defend their discipline as a "hard" science distinct from cosmology as mere imaginary. A relatively recent subdiscipline within string theory seeks to synthesize cosmology with its own microscopic, Planck-scale formalisms.

14. Heliocentrism may well be easy enough for a lay person to validate experimentally with some guidance and perseverance, but what about the existence of atoms, let alone strings?

15. Consider, for example, that while a matter of speculation since Galileo's telescopic observations of the Milky Way and what were then called nebulae, galaxies did not become a definitive feature of the known universe until the 1920s, through the work of Edwin Hubble. Before then, in terms of an imaginary, we lived in a world without galaxies. Alexander Koyré writes of the transformation in cosmology (in both the cultural and scientific senses) that occurred during the early modern period—from what he calls a "closed world" to an "infinite universe" (2).

16. Lakoff and Johnson describe how a relatively finite set of images form a kind of *doxa* that, in effect, limits imaginative freedom: "The study of spatial-relations concepts within cognitive linguistics has revealed that there is a relatively small collection of primitive image schemas that structure systems of spatial relations in the world's languages. Here are some examples . . . : part-whole, center-periphery, link, cycle, iteration, contact, adjacency, forced motion (e.g., pushing, pulling, propelling), support, balance, straight-curved, and near-far. Orientations also used in the spatial-relations systems of the world's languages include vertical orientation, horizontal orientation, and front-back orientation" (*Philosophy* 35).

17. Needless to say, Feyerabend takes a dim view of scientism for non-scientists: "The lesson I draw from this sequence of events is that a uniform 'scientific view of the world' may be useful for people doing science—it gives them motivation without tying them down. It is like a flag. Through presenting a single pattern it makes people do many different things. However, it is a disaster for outsiders (philosophers, fly-by-night mystics, prophets of the New Age). It suggests to them the most narrowminded religious commitment and encourages a similar narrowmindedness on their part" (250).

18. Clarke conflates technology with science (technoscience) in order to emphasize not only their interdependence, but the impossibility of delineating the boundary between the two as multiplicities.

19. Galileo asserts that "the Universe, which stands continually open to our gaze, cannot be understood unless one first learns to comprehend the language and read the letters in which it is composed. It is written in the language of mathematics" (237–38). This ascription is not merely a poetic caprice; it is important to acknowledge that mathematics has been integral to many, if not most, of the scientific and technological advances of the last few centuries.

20. Named after the mathematicians who first described them—Eugenio Calabi and Shing-Tung Yau —Calabi-Yau manifolds are particular multidimensional spaces with special features that are used in certain versions of superstring theory. In topology, orientifolds and orbifolds are, roughly speaking, similar yet distinct ways of structuring manifolds.

21. "From [1995] on, the name 'superstring theory' becomes something of a misnomer, since those working in this field now feel that they are studying bits of a

much larger theory that contains not only strings but higher-dimensional *p*-branes. Besides M-theory, this theory has also been sometimes called 'the theory formerly known as strings'" (Woit, *Not Even Wrong* 156).

22. For a popular account, see Susskind, *The Cosmic Landscape* 89–109.

23. Woit writes: "To really understand superstring theory, one should first study quantum field theory, and in itself this is a very demanding task. Typically, graduate students take a course in quantum field theory during their second year of graduate school, in which case they can't even begin work on superstring theory until their third year. This leaves little time to master the subject and get some new results about it in the standard four- to five-year length of the PhD program. Young superstring theorists often gain their degrees having only become really familiar with a small part of the subject, with true expertise requiring many, many years to achieve" (*Not Even Wrong* 200).

24. The holographic principle is a property in some versions of string theory where the description of the volume of a given space is equivalent to a description of the boundary of that space. One version of this is called *Anti-de-Sitter Space/Conformal Field Theory*, or *AdS/CFT*. See Maldacena, "The Large N Limit of Superconformal Field Theories and Supergravity"; or, for popular accounts, Penrose 920–22 and Woit, *Not Even Wrong* 156–57.

25. An *ansätz*, from the German meaning "attempt," is an educated guess at the solution to a problem.

26. For example, Zwiebach writes: "Einstein's equations of general relativity are elegant and geometrical. They embody the conceptual foundation of the theory and feel completely up to the task of describing gravitation. No similar equations are known for string theory" (10–11). Greene asserts: "But most researchers feel that our current formulation of string theory still lacks the kind of core principle we find at the heart of other major advances. Special relativity has the constancy of the speed of light. General relativity has the equivalence principle. Quantum mechanics has the uncertainty principle. String theorists continue to grope for an analogous principle that would capture the theory's essence as completely" (*Fabric* 376).

27. For string theorists, a statement such as "the world is made of strings," or "all quantum particles are variations on a fundamental string" does not qualify as a principle, although their rationale for this is somewhat obscure.

28. One example of an image from the general relativity imaginary is the bending and warping of the space-time fabric.

29. Lakoff and Núñez contend that the image of infinity consists of the following: "processes that go on indefinitely are conceptualized as having an end and ultimate result, an infinite thing" (158). In effect, an "infinite thing" becomes a container that holds an indefinite process.

Chapter 3

1. Note that partons, mesons, and hadrons are not on par in the standard model's taxonomy. Mesons are a kind of hadron consisting of one quark and one antiquark. Partons now are recognized as either quarks or gluons.

2. According to the *OED*, it was Michael Faraday, from a suggestion by his col-

league William Whewell, who first used the suffix *-on* in a physics context, with his neologism *ion*. The *-on* derives from the Greek *–ov*, and its grammatical function is to terminate neuter nouns.

3. In 1939, H. J. Bhabha wrote in *Nature:* "It is felt that the 'tr' in this word is redundant, since it does not belong to the Greek root 'meso' for middle; the 'tr' in neutron and electron belong, of course, to the roots 'neutr' and 'electra.' It would therefore be more logical and also shorter to call the new particle a meson instead of a mesotron" (276).

4. These haphazardly applied naming conventions remind me of a similar capriciousness in the naming of discovered lands in the New World during the early modern period.

5. In her popularization *Warped Passages*, Randall asserts: "We will now see that branes are more than just a location; they are objects in their own right. Branes are like membranes, and, like membranes, they are real things" (305).

6. Consider how different an apple would appear to a dog, an ant, or a bacterium.

7. Lakoff and Johnson write: "We experience space as structured by image schemas (as having bounded regions, paths, centers and peripheries, objects with fronts and backs, regions above, below, and beside things). Yet we now know that space in itself has no such structure. The topographic maps of the visual field, the orientation-sensitive cells, and other highly structured neural systems in our brains not only create image-schematic concepts for us but also create the experience of space as structured according to those image schemas" (*Philosophy*, 508–9).

8. Bruce Hunt's essay, "Lines of Force, Swirls of Ether," in *From Energy to Information*, investigates the mechanical models that nineteenth-century physicists, including Maxwell, used to represent the luminiferous ether, and the subsequent mathematical abstraction of these models that accompanied the rise of electromagnetic theory. Such mathematical abstractions rendered the mechanical models obsolete. Hunt's essay hints at the material and imaginative processes by which a scientific concept transforms from the concrete to the abstract.

9. Note that this is "force" in the ordinary sense of someone or something pushing or pulling something else, not "force" in any of its technical definitions within physics.

10. "In distant miniatures," Bachelard writes, "disparate things are reconciled. They then offer themselves for our 'possession,' while denying the distance that created them. We possess from afar, and how peacefully!" (*Poetics* 172)

11. As noted in chapter 1, with the emergence of M-theory in the mid-1990s, the brane has come to supersede the string's centrality as fundamental object. Lisa Randall notes that "the theorist Michael Duff facetiously refers to 'string theory' as 'the theory formerly known as strings'" (*Warped* 304).

12. Veneziano, "Construction of a Crossing-Symmetric, Regge-Behaved Amplitude for Linearly Rising Trajectories."

13. The symbol π here refers to *pion*.

14. Recall Einstein's famous equation $E = mc^2$, where E, energy, and m, mass, share a fundamental equivalence.

15. In *The Cosmic Landscape*, Susskind defines a harmonic oscillator as "anything that can vibrate or swing back and forth with a periodic (repeating) motion" (205).

16. Recall that a meson is a kind of hadron that is composed of one quark and one antiquark.

17. Susskind anticipates this issue when he writes shortly thereafter: "The reader may now begin to wonder how a non-relativistic 4-dimensional Schrödinger equation can apply to the internal structure of a hadron. . . . The final rule about 4-dimensional quantum mechanics which we adopt is that we assume that our equations are valid in the Wick-rotated world . . . At the end of the calculation we shall have to perform the familiar analytic continuation to get back the real Minkowski world" (464). Wick rotations, named after the Italian physicist Gian-Carlo Wick, allow theorists, generally speaking, to solve solutions in quantum theory by replacing real number variables with imaginary number variables. Significantly, these mathematical procedures allow theorists to transform one imagined space into another, one which enjoys a "real" status.

18. In quantum theory, these operators provide for the perpetual creation and annihilation of virtual particles.

19. Note that, technically speaking, the concept of spin, or angular momentum, is an abstraction. In quantum theory, since fundamental particles have no extent in space, they cannot literally spin, like a ball.

20. According to Andrew Pickering, it was Yoichiro Nambu who first made the association between the image of the string and the behavior of hadrons. Nambu was an expert in solid-state physics and had borrowed the concept of the string from his study of superconductors, where magnetic flux was theorized to get trapped inside the superconductors in string-like vertices (*Constructing* 275n22). But it was the positive reception of Susskind's paper by the theoretical physics community that gave the term its staying power.

21. Elsewhere in *The Cosmic Landscape*, Susskind offers yet another alternative, supplementary scenario: "The String Theory of hadrons can be pictured in just this way. The world sheets, tubes, and Y-joints are really just very complicated Feynman diagrams involving quarks and a very large number of gluons. When you look at the world sheet from a distance, it appears smooth. But under a microscope it looks like a 'fishnet' or 'basketball-net'" (216).

22. In a personal interview, the physicist Don Petcher playfully suggested the name "spring theory," which has a certain aptness in that it emphasizes the mathematically-expressed mechanical setup that Susskind actually uses. In contrast to the string, the spring would have its own cache of imaginative associations.

23. Lakoff further argues that "event categories and other abstract categories are structured . . . on the basis of structures from the realm of physical experience" (48).

24. Zwiebach writes: "Calabi-Yau compactification of heterotic strings gave the first string models with semi-realistic particle physics" (344). It is semi-realistic because the particle spectrum it generates does not reproduce all three gauge groups—for electromagnetism and the strong and weak nuclear forces—of the standard model with the necessary particle attributes.

25. The word *heterosis* has an interesting history. The 1864 edition of *Webster's Unabridged Dictionary* defines it as a term from Rhetoric meaning "a figure of speech by which one form of a noun, verb, or pronoun, and the like, is used for another"; the *OED* reports that in the early part of the twentieth century it was

used in zoology to mean segmentation, and later in the century, in genetics to mean hybrid vigor through cross-breeding. This would seem to be another instance of a word that has made the journey in the sciences away from the concrete to the abstract, from the human-scaled organic to the remotely-scaled inanimate.

26. For example, consider a video game where a spaceship disappears on the right side of the screen only to instantaneously reappear on the left side of the screen. While the physical space of the game appears flat on the monitor, the abstract mathematical space on which the game is based is that of a torus, where the right seam, so to speak, of the vertical dimension is continuous with the left seam. The *OED* defines the torus as "a surface or solid conceived of as generated by the circular motion of a circle about an axis outside itself but lying in its plane; also, any body topologically equivalent to this, having one hole in it but not necessarily circular in form or cross-section."

27. A *soliton* is a particular kind of quantum particle that propagates as a single wave. An *isometry* is the transformation of one metric space into another that preserves the distances between points.

28. Lakoff argues that "meaningful thought is not merely the manipulation of abstract symbols that are meaningless in themselves and get their meaning only by virtue of correspondences to things in the world. Reason is not abstract and disembodied, a matter of instantiating some transcendental rationality. The mind is thus not simply a 'mirror of nature,' and concepts are not merely 'internal representations of external reality'" (370).

29. While beyond the scope of this book, the "constructivist" social systems theory of Niklas Luhmann posits an epistemology that complements, in many respects, my notion of "triangulation" here. They both suggest a resolution to the "subject/object" aporia by means of a "de-ontologization" of scientific realism ("Cognitive" 131–32). See, for example, Luhmann, "The Cognitive Program of Constructivism and the Reality That Remains Unknown."

30. Newton's contemporaneous critics certainly did not find the notion of action-at-a-distance in his theory of gravitation to be coherent or convincing. They argued that gravity requires a medium of propagation.

31. The "lore" they presumably refer to here is that we only live in four non-compact dimensions since such a belief is emphatically confirmed by the common-sense experience of our bodies. We move in three dimensions of space. Time feels dimension-like in that it progresses linearly from past to future. The lore is that any other dimensions then could not be non-compact because we do not experience them "directly."

32. In their paper "Brane World," theorists Zurab Kakushadze and S. H. Henry Tye introduce the expression "brane world" to describe a "scenario, where the Standard Model gauge and matter fields live inside some branes while gravitons live in the bulk" (13).

33. In *Warped Passages*, Randall defines *Kaluza-Klein particles* as "particles that originate in extra dimensions, but appear to us as extra particles in our four-dimensional spacetime . . . for each particle that we know about, there should be many KK particles with the same charge, each with a different mass" (353, 354). In keeping with the shifting from object-event to location-event frames, Randall and Sundrum alternatively refer to them as "KK modes."

34. What the writers suggest here is that with only a small, almost negligible adjustment in values, calculations using their braneworld model of gravity will produce the same results as general relativity. This is a correction that contemporary technologies are only beginning to be sensitive enough to detect. In effect, both models, their braneworld one and general relativity, have the same predictive power.

35. Brian Greene muses briefly on this conundrum: "Does it mean that on a microscopic level the universe operates in ways so obscure and unfamiliar that the human mind, evolved over eons to cope with phenomena on familiar everyday scales, is unable to fully grasp 'what really goes on'? Or, might it be that through historical accident physicists have constructed an extremely awkward formulation of quantum mechanics that, although quantitatively successful, obfuscates the true nature of reality? No one knows" (*Elegant* 87).

36. Le Doeuff argues that "images in philosophical texts regularly turn out to have been taken over from precise earlier sources, making their study into part of a history of learning's fabulous motifs" (9).

37. I want to emphasize here that the imaginaries at stake in string theory technical exposition echo heroic romance rather than Romanticism. Generally speaking, within the Romantic tradition, the natural world is imaginatively framed as terrifyingly powerful. As a consequence, in encounters with the natural world, the appropriate emotional response is couched in terms of the sublime. String theory technical exposition does not seem to bring such a deferential respect, however ambivalent in Romanticism, to its imagined encounters with the microcosm.

38. See, for example, Huggan and Tiffin; Thurtle; Thurtle and Mitchell; Wald; Tabbi; Wutz and Tabbi.

39. See, for instance, Zunshine.

40. This use of the first person plural may also accrue a certain added authority in imitation of the royal 'we.' It may also reflect an emphasis in theoretical physics on group consensus—that the plural voice has the most gravity. Both of these interpretations emphasize the subordination of an individual identity to a higher order.

41. Even the misprint "Le" speaks to a mood, however inadvertent, of hasty breathlessness, of a eureka moment.

Chapter 4

1. The first monographic popularization of string theory is Kaku's *Beyond Einstein*, a best seller coauthored with Jennifer Trainer and published in 1987. Also in 1987, BBC Radio broadcast a series of interviews with theorists leading the "first superstring revolution," as it has been retroactively called. The following year, Cambridge University Press published a collection of edited transcripts of those BBC interviews, entitled *Superstrings: A Theory of Everything?*, edited by P. C. W. Davies and Julian Brown. In the wake of the commercial success of *Hyperspace*, Anchor Books republished an updated version of *Beyond Einstein* in 1995.

2. While Kaku has published more than seventy technical articles and enjoys, as of this writing, an h-index of 23, most theorists regard his early contributions to string field theory as the most significant. The h-index is a ranking within a given

scientific discipline that measures the frequency of citation. A higher number represents more citations. Ed Witten, who first introduced M-theory to the field, has the highest h-index of any living physicist: 142.

3. These include *Visions: How Science Will Revolutionize the 21st Century*; *Parallel Worlds: A Journey Through Creation, Higher Dimensions, and the Future of the Cosmos*, which the BBC made into an eponymous one-hour programme; *Physics of the Impossible: A Scientific Exploration into the World of Phasers, Force Fields, Teleportation, and Time Travel*; and most recently, *Physics of the Future: How Science Will Shape Human Destiny and Our Daily Lives by the Year 2100*.

4. Sponsored by the Royal Society's Committee for the Public Understanding of Science, since 2007, the Aventis Science Book Prize has been known as the *Royal Society Prize for Science Books*.

5. In some important versions of superstring theory, the extra six space-time dimensions are compactified into Calabi-Yau manifolds in order to more accurately reproduce the particle spectrum of the standard model. Greene and Plesser found that certain Calabi-Yau manifolds, while geometrically distinct, could be paired with other Calabi-Yau manifolds because they were physically equivalent. By choosing the one in the pair that is less mathematically complex, this simplifies the work theorists need to do when formulating superstring models. (*Elegant* 256–59)

6. Space-tearing flop transitions are a mathematical means for transforming one Calabi-Yau manifold into its mirror pair by an abstracted "pinching" and "puncturing" of the space-time fabric. They found that this tearing of space-time is not, in fact, disastrous—a longstanding worry for theorists. (*Elegant* 266–82)

7. In *Warped Passages*, Randall defines model building thus: "what physicists call the search for theories that might underlie current observations" (8). This definition would seem to emphasize inductive reasoning, as opposed to theorizing per se, which tends to be deductive (i.e., moving from the general to the particular). But as the likes of Feyerabend certainly would attest, the relationship between model building and theorizing is more complicated than that.

8. To summarize, physicists have yet to observe through experimentation the supersymmetric partners (or sparticles) of standard model particles. In Randall and Sundrum's sequestered symmetry breaking model, sparticles become sequestered away from the braneworld that is our universe, but exist in the enveloping bulk, which would explain why we cannot observe them. The hierarchy problem is that in order to probe smaller and smaller subatomic scales, accelerators require ever more energy. In keeping with Einstein's $E = mc^2$, more energy means a larger mass. Accordingly, Planck scale objects would be relatively massive (e.g., a black hole). It is a mystery to theorists why there is such a large gap in quantity between the Planck scale mass and the masses of ordinary standard model subatomic particles. Randall writes: "Another way of phrasing the hierarchy problem is to ask why the Planck scale mass is so huge—why is it ten million billion times higher than the masses relevant to particle physics scales, all of which are less than a few hundred GeV?" (250). The relevant technical articles are Randall and Sundrum, "Out Of This World Supersymmetry Breaking," and "A Large Mass Hierarchy from a Small Extra Dimension."

9. As of August 2012, SPIRES, the online high energy physics bibliography, reports that "An Alternative to Compactification" has been cited 4,590 times; "Out

of this World," 1,096 times; and "A Large Mass Hierarchy," 5,449 times. For a community of a couple thousand working theorists, these figures are considerable.

10. To reiterate, one may roughly summarize this work as that which attempts to use M-theory influenced formulations of branes and non-compact extra dimensions to explore new cosmological models, models that incorporate braneworlds into a multidimensional metaverse or bulk.

11. There is a widely circulating bit of apocrypha that when writing *A Brief History of Time*, an editor warned Stephen Hawking that the inclusion of any equation would halve sales.

12. Greene and Randall tuck away a few simplified mathematical formulae in the endnotes. Moreover, these formulae are isolated from the full context of the original mathematical argument, rendering their significance all the more obscure.

13. To reiterate, I have adopted the notion of object- and location-event schema from Lakoff and Johnson, *Philosophy* (177–99).

14. Recall the discussion in the previous chapter on the particular definition of string where it stands metonymically for that which it bears or contains, whether associated attributes or information.

15. In *The Elegant Universe*, Greene elevates three personages to his pantheon: Isaac Newton, Albert Einstein, and Ed Witten (the originator of M-theory): "Newton's was such a monumental intellect that, for example, when he found that the mathematics required for some of his investigations did not exist, he invented it. Nearly three centuries would pass before the world would host a comparable scientific genius" (54); "By offering the explanation for the expansion of the universe, Einstein achieved one of the greatest intellectual feats of all time" (82); "Edward Witten's razor-sharp intellect is clothed in a soft-spoken demeanor that often has a wry, almost ironic, edge. He is widely regarded as Einstein's successor in the role of the world's greatest living physicist. Some would go even further and describe him as the greatest physicist of all time. He has an insatiable appetite for cutting-edge physics problems and he wields tremendous influence in setting the direction of research in string theory" (274).

16. Recall from the introduction that string theory originates in Veneziano's adaptation of the Beta function—an antique formula attributed to Leonhard Euler—to the problem of hadrons. In *Nature's Numbers*, Ian Stewarts reports that in the first half of the eighteenth century, the efforts of leading European mathematicians, using the tool of calculus they inherited from their immediate predecessors, Newton and Leibniz, were almost universally focused on developing a formalism to describe the dynamics of the violin string. Stewart writes: "In 1748 . . . the prolific Swiss mathematician Leonhard Euler worked out the 'wave equation' for a string. In the spirit of Isaac Newton, this is a differential equation that governs the rate of change of the shape of the string. . . . Not only did Euler formulate the wave equation: he solved it" (64).

17. This is how the string is sometimes represented in graphic illustrations and video representations.

18. Recall that Kaku defines a *field* as "a collection of numbers defined at every point in space that completely describes a force at that point" (*Hyperspace* 25). He observes that "Faraday got this concept when he thought of a 'field' ploughed by a farmer" (25).

19. It ought to be noted that the ancients made use of simple yet versatile instruments such as the sundial and the gnomon–a straight stick of unit length planted in the ground.

20. Lakoff and Johnson explain that a "folk theory" is "a basic explanatory model shared by most . . . We intend nothing at all derogatory by describing this knowledge as a 'folk theory.' On the contrary, it is just such models that make up a culture's common sense" (352). Folk theories can form the basis of both commonsense knowledge and higher-order philosophical or scientific theories. Faculty Psychology relies, they argue, on a "Society of Mind" metaphor where "each capacity of the mind is conceptualized" as a distinct person-like entity. These faculties are: "Perception, Imagination, Feeling, Will, Understanding, Memory, and Reason" (410).

21. Lakoff and Johnson write: "Cognitive science and neuroscience suggest that the world as we *know* it contains *no* primary qualities . . . because the qualities of things as we can experience them depend crucially on our neural makeup, our bodily interactions with them, and our purposes and interests. For real human beings, the only realism is an embodied realism" (26, emphasis in original). Consider how even a relatively straightforward concept such as *star* has changed over the centuries.

22. The idea of a space being a "region" strikes me as rather tautological. As discussed in chapter 3, it would seem to depend on the ready, and archaic, association between the idea of space and that of geographic place.

23. There are numerous examples of extra-dimensional "informational" spaces in theoretical physics; for example, configuration and phase spaces. These are spaces that describe more than just the position of an object in "real" space; they include within the multidimensional space supplemental information such as momentum or spin. In chapter 3, I called them "attribute" spaces to emphasize the relationship between these spaces and the attributes of the phenomena that such spaces describe.

24. To speak of "features" is to also invoke a language of consumerism where products are purchased on the basis of features, what they do, and benefits—the feelings associated with the results of the having done.

25. As noted above, this occurs in chapters 10 and 11 where he narrates the story of his work on "mirror symmetry" and "space-tearing flop transitions," respectively. Like Kaku, these anecdotes are situated within a history of string theory, thereby emphasizing their relative importance.

26. Thomas Lessl writes of the "priestly voice" in the popularizations of Carl Sagan.

27. Kaku makes almost the exact same declaration: "It seems almost incomprehensible that we, as intelligent apes on the third planet of a minor star in a minor galaxy, would be able to reconstruct the history of our universe going back almost to the instant of its birth, where temperatures and pressures exceeded anything ever found in our solar system. Yet the quantum theory of the weak, electromagnetic, and strong interactions reveals this picture to us" (214).

28. While fertile ground for investigation, it is beyond the scope of this book to examine specifically the gender politics of the imaginary that these three popularizations each presents.

Chapter 5

1. I equate Le Doeuff's term "enterprise" with what I call "discourse."

2. They are what Roland Barthes would call "readerly" texts. Kaja Silverman explains: "The readerly text . . . attempts to conceal all traces of itself as a factory within which a particular social reality is produced through standard representations and dominant signifying practices"; the "writerly" text "exhumes the cultural voices or codes responsible for the latter's enunciation and in the process it discovers multiplicity instead of consistency and signifying flux instead of stable meaning" (240, 246).

3. As mentioned in chapter 2, Henderson suggests something comparable to "imaginative parataxis" when she writes that "not only was it a popular fascination, but the idea of the fourth dimension as a place or as a temporal means of reaching another era provided a position from which to comment on contemporary society" (33). In *Energy Forms*, Clarke argues that "scientific allegory" (discussed at the end of chapter 3) represents a move from "momentary positional juxtaposition to sustained spatiotemporal structure" (26). My concern here is not so much the extent to which such imaginative juxtapositions are sustained, but rather, the imaginative leap at stake between juxtaposed epistemological domains. Generally speaking, the wider the gap, as with Planck and quotidian human scales, the more jarring—and telling—the parataxis.

4. Science plays obviously are one particular form of imaginative writing.

5. *String Fever* premiered at The Ensemble Studio Theatre, New York, on February 26, 2003.

6. On the dedication page, Reingold gives a "special thanks to Brian Greene and his wonderful book *The Elegant Universe*." Anyone with a passing familiarity with that book and its author will find it difficult not to imagine that the character Frank is based loosely on Greene himself.

7. "Like an infinitely thin rubber band, each particle contains a vibrating, oscillating, dancing filament that physicists, lacking Gell-Mann's literary flair, have named a *string*" (*Elegant* 14, emphasis in original).

8. Or triply derived, if one considers images in technical exposition already a doubling of mathematical concept.

9. *Humble Boy* premiered at the Royal National Theatre, London, in August 2001.

10. Presumably, Greene chose the word "object" for good reason; a "one-dimensional *line*" is tautological.

11. "According to string theory, the elementary ingredients of the universe are *not* point particles. Rather, they are tiny, one-dimensional filaments somewhat like infinitely thin rubber bands, vibrating to and fro" (*Elegant* 136, emphasis in original).

12. A cosmic string is a hypothesized one dimensional topological defect in space-time that is extremely dense and may be responsible for the clumping of matter shortly after the Big Bang.

13. Traweek observes that "in principle, theorists and experimentalists at laboratories must work closely together, but they usually exhibit a strong wariness to each other. . . . Theorists tend to have more status at a younger age; they work

in short-lived collaborations with one or two other theorists; they 'do physics' at blackboards in their offices" (111). Experimentalists, on the other hand, tend to work in larger groups with colliders and accelerators testing theory against evidence.

14. See Strauss; and Collins.

15. But, as Paul Collins points out in his review of the show on *Slate.com*, the most interesting aspect of the show's characterization of Leonard as a string theorist is in the ways in which his behavior suggests the symptoms of Asperger's syndrome. According to Collins, the show becomes "a meditation on how bright people work with the absurdly mismatched abilities that they've been given" in the decidedly domestic context of the American-style sitcom.

16. We also saw the epistemological conceit of reality-as-depth at the end of the technical article "Heterotic String Theory (I)."

17. According to his university webpage, Benford's research program "unites theoretical studies with a parallel experimental program in radiation processes of relativistic electron streams in plasma. . . . These experiments and coupled theory apply to galactic jets, quasars, and pulsars."

18. When Benford writes that "they opened the portal into the looking-glass-like Counter system," the two astronauts' voyage through the "Q dimension" to "Counter-Earth" is explicitly compared to Lewis Carroll's *Alice in Wonderland* (313). This is another evocation of an authority—in this case, an authority that permits imaginative license.

19. This is a notion, techno-scientific power, that I discuss at length later in the chapter.

20. Kaku published a book that focuses specifically on this theme: *Physics of the Impossible: A Scientific Exploration into the World of Phasers, Force Fields, Teleportation, and Time Travel.* Other notable examples of this form of sci-tech speculation include Lawrence Krauss's *The Physics of* Star Trek and *Beyond* Star Trek: *From Alien Invasions to the End of Time.*

21. Greene also speaks of scientific knowledge in a romantic, if not quasi-religious register: "Although we are technologically bound to the earth and its immediate neighbors in the solar system, through the power of thought and experiment we have probed the far reaches of both inner and outer space. During the last hundred years in particular, the collective effort of numerous physicists has revealed some of nature's best-kept secrets. And once revealed, these explanatory gems have opened vistas on a world we thought we knew, but whose splendor we had not even come close to imagining" (286).

22. Greene devotes only two pages to wormholes in *The Elegant Universe* (265–66), but in a follow-up popularization entitled *The Fabric of the Cosmos*, he spends over thirty pages contemplating them and the possibilities of time travel (437–70).

23. Recall in the previous chapter the passage in *Hyperspace* that likens strings to DNA (156).

24. As an example of the cringe-worthy prose on display in this novel, take the narrator's introduction of the character Monique: "What he noticed first was how tall she was, a full head taller than the wizened chairman of the physics department, who introduced Monique as 'the brightest young student I've had the pleasure of working with.' David wondered if perhaps the old man had grown overly fond of her, because in addition to being tall, this woman was beautiful. Her face was like

one of the ancient portraits of Athena, the Greek goddess of wisdom, but instead of a helmet she wore a crown of intricately woven cornrows, and her skin was the color of a Kahlúa and cream. A long dress made from yellow-and-red Kente cloth draped her shoulders, and several gold bracelets hung from each of her brown arms. In the drabness of Jadwin Hall she blazed like a particle shower" (66).

25. Roughly speaking, the ekpyrotic model (from the Greek meaning "out of fire") argues that our universe originates not in a Big Bang but in a Big Splat. Two braneworlds positioned closely together in a bulk cyclically collide, forming the universe we inhabit. See Steinhardt and Turok 121–66.

26. "S-Bomb" appears in *Riffing on Strings: Creative Writing Inspired by String Theory*. It is a collection of new and previously published works of nonfiction, fiction, poetry, and drama that engage with string theory in some way, and that I coedited.

27. "Singularity" is a jargon term from cosmology that generally refers to a space-time event of infinite density, such as a black hole or the Big Bang. Here it apparently means something more mundane—that all things in the cosmos are joined as one.

28. Rucker's "Guadalupe and Hieronymus Bosch," originally published in *Interzone* 200 (September 2005), was reprinted in *Year's Best SF 11* (89–105); and Malcohn's "Arachne" was first published in *Aboriginal Science Fiction* (Nov./Dec. 1988) and reappeared in *Riffing on Strings* (93–110). Page numbers refer to the latter in both cases.

29. Hilbert space, named after the early twentieth-century German (the English-sounding surname notwithstanding) mathematician David Hilbert, is an infinite-dimension function space, an abstract space that measures distances and angles (vectors). One relevant application of Hilbert space is in modeling the harmonics of vibrating branes.

30. *Strings* premiered at the 78th Street Theatre Lab, New York, in 2006. Other plays I have encountered that engage with string theory include Susanna Speier, *Calabi-Yau*, which premiered at the HERE Arts Center, New York, in 2001; Simon McBurney, *A Disappearing Number*, which premiered at the Theatre Royal Plymouth, England, in 2007; and Drucilla Cornell, *Background Interference* (unpublished manuscript). I do not consider them here as their engagement with string theory is no more nuanced than the examples already provided.

31. To reiterate, the ekpyrotic model argues that our universe originates not in a Big Bang but in a Big Splat. Two braneworlds positioned closely together in a bulk cyclically collide, forming the universe as we know it.

32. Strictly speaking, a proton, neutron, and electron together would form deuterium, a stable isotope of hydrogen.

33. There is another possible motivating factor for June's infidelity that I have not pursued. Through the course of act 1, we learn that she feels guilty for not having died with her son on September 11, 2001, when she failed to meet him for a date at Windows on the World at the Twin Towers in Manhattan, because she had been delayed on the subway. This would be a psychoanalytic reading that suggests that the taking of a lover was, in some sense, a way of replacing the lost intimacy with the son, and, through its transgressivity, a way of confirming her own moral shortcomings. But I find the involvement of 9/11 imagery in *Strings* somewhat gratuitous—a perhaps too obvious effort to be all-inclusively topical. The conceit

of the loss of a child due to an unforeseen calamity could have had a potent bearing on the psychological makeup of the characters without the evocation of 9/11, which introduces other thematic complications not fully explored.

34. Page numbers refer to the *Riffing on Strings* reprint. "String Theory Sutra" originally appeared in Brenda Hillman, *Pieces of Air in the Epic* (Middletown, CT: Wesleyan University Press, 2005).

35. Or, in other words, the implied author, whom I will refer to as Hillman.

36. Ginsberg's poem, in turn, references William Blake's poem "Ah! Sun-flower," in which the sunflower seeks "after that sweet golden clime/Where the traveller's journey is done," a place contrasted with "graves . . . shrouded in snow" (39).

37. For example, as in Fritjof Capra's *The Tao of Physics: An Exploration of the Parallels between Modern Physics and Eastern Mysticism.*

38. See the graphic illustration of a Calabi-Yau space (figure 8.9) in *The Elegant Universe* (207). In an endnote, Greene writes: "This illustration is courtesy of Andrew Hanson of Indiana University, and was made using the *Mathematica* 3-D graphing package" (401n9).

Chapter 6

1. In *Three Roads to Quantum Gravity*, Lee Smolin defines background independence thus: "Neither space nor time has any existence outside the system of evolving relationships that comprises the universe. Physicists refer to this feature of general relativity as *background independence.* By this we mean that there is no fixed background, or stage, that remains fixed for all time" (24–25, emphasis in original).

2. Furthermore, as far as the spectacle of its popularization goes, there is something exquisitely Catholic about string theory—theorists speak in a language, the language of, among others, differential calculus and Riemannian geometry, that is opaque to the lay audience, much in the same way that medieval plebeians could consume the Latin scriptures only as mediated by and through priests. Its critics even complain of string theory's "scholasticism," divorced as it currently is from experiment.

3. While British, European, and Asian theorists have clearly made significant contributions to string theory, many historians of education would argue that since World War II, as American higher education has emerged as the dominant system, other countries have worked to emulate the U.S., gradually shifting policy control from central government ministries to the institutions themselves in an effort to become more responsive to the higher education market. Nevertheless, the European high energy physics community often perceives itself in a fierce rivalry with their Anglo-American counterparts. See, for example, Trow, "Comparative Perspectives on British and American Higher Education," 280–99, and Wittrock, "The Modern University: The Three Transformations," 303–62, in Rothblatt and Wittrock.

4. See Geiger, "Research, Graduate Education, and the Ecology of American Universities: An Interpretive History" in Rothblatt and Wittrock 258.

5. See Traweek, 2; and Graham and Diamond 8.

Bibliography

Abbas, Niran, ed. *Mapping Michel Serres.* Ann Arbor: U of Michigan P, 2005. Print.

Allen, Graham. *Intertextuality.* London: Routledge, 2000. Print.

Alpert, Mark. *Final Theory.* New York: Simon & Schuster, 2008. Print.

Anderson, Philip. Interview by Science Desk. "God (or Not), Physics and, of Course, Love: Scientists Take a Leap." *New York Times* 4 Jan. 2005, F3. Web. 1 June 2009.

Aronowitz, Stanley. "Alan Sokal's 'Transgression.'" *Dissent* 44.1 (1997): 107–10. Print.

Assad, Maria. *Reading with Michel Serres: An Encounter with Time.* Albany: State U of NY P, 1999. Print.

Bachelard, Gaston. *The Formation of the Scientific Mind: A Contribution to a Psychoanalysis of Objective Knowledge.* Trans. Mary McAllester Jones. Manchester, UK: Clinamen, 2002. Print.

Bachelard, Gaston. *The New Scientific Spirit.* Trans. Arthur Goldhammer. Boston, MA: Beacon Press, 1984. Print.

Bachelard, Gaston. *The Poetics of Reverie: Childhood, Language, and the Cosmos.* Trans. Daniel Russell. Boston: Beacon Press, 1969. Print.

Bachelard, Gaston. *The Poetics of Space.* Trans. Maria Jolas. Boston: Beacon Press, 1969. Print.

Baez, John. "Physics Bitten by Reverse Alan Sokal Hoax?" *Sci.physics.research Group.* Google, 24 Nov. 2002. Web. 10 Sept. 2009.

Baigrie, Brian S., ed. *Picturing Knowledge: Historical and Philosophical Problems Concerning the Use of Art in Science.* Toronto: U of Toronto P, 1996. Print.

Banks, Tom, Nathan Seiberg, and Stephen Shenker. "Branes from Matrices." *Nuclear Physics* B490 (1997): 91–106. *ArXiv.org.* Web. 1 June 2009. hep-th/9612157v2.

Barbour, Julian. *The End of Time: The Next Revolution in Physics.* Oxford: Oxford UP, 1999. Print.

Barnes, Barry, David Bloor, and John Henry. *Scientific Knowledge: A Sociological Analysis.* London: Athlone, 1996. Print.

Barr, Marleen S., ed. *Envisioning the Future: Science Fiction and the Next Millennium.* Middletown, CT: Wesleyan UP, 2003. Print.

Barrow, John D. *Theories of Everything: The Quest for the Ultimate Explanation.* New York: Oxford UP, 1991. Print.

Barthes, Roland. *S/Z: An Essay.* Trans. Richard Miller. New York: Hill and Wang, 1974. Print.
Baxter, Stephen. *Ring: Riding the Superstrings to the Edge of the Universe.* New York: Eos, 1994. Print.
Becker, Katrin, Melanie Becker, and John H. Schwarz. *String Theory and M-Theory: A Modern Introduction.* Cambridge: Cambridge UP, 2007. Print.
Beer, Gillian. *Darwin's Plots: Evolutionary Narrative in Darwin, George Eliot, and Nineteenth-Century Fiction.* 2nd ed. Cambridge: Cambridge UP, 2000. Print.
Beer, Gillian. *Open Fields: Science in Cultural Encounter.* Oxford: Clarendon, 1996. Print.
Benford, Gregory. "Faculty: Gregory Benford." *Physics.uci.edu.* Dept. of Physics and Astronomy, U of California, Irvine, 22 June 2008. Web. 24 July 2008.
Benford, Gregory. "On the Brane." Hartwell and Cramer 310–29. Print.
Best, Steven. "Chaos and Entropy: Metaphors in Postmodern Science and Social Theory." *Science as Culture* 11.2 (1991): 188–226. Print.
Best, Steven, and Douglas Kellner. *The Postmodern Adventure: Science, Technology, and Cultural Studies at the Third Millennium.* New York: Guilford, 2001. Print.
Beyer, Kirsten. *Fusion: Star Trek Voyager String Theory, Book 2.* New York: Simon & Schuster, 2005. Print.
Bhabha, H. J. "The Fundamental Length Introduced by the Theory of the Mesotron (Meson)." *Nature* 143.3616 (1939): 276–77. Print.
Blake, William. *Songs of Innocence and Songs of Experience.* Mineola, NY: Dover, 1992. Print.
Bohnenkamp, Dennis. "Post-Einsteinian Physics and Literature: Toward a New Poetics." *Mosaic* 22 (1989): 19–30. Print.
Borges, Jorge Luis. "The Analytical Language of John Wilkins." *Other Inquisitions: 1937–1952.* Trans. Ruth L.C. Simms. Austin: U of Texas P, 1964. 101–5. Print.
Brockman, John. *The Third Culture.* New York: Simon & Schuster, 1995. Print.
Broderick, Damien. *Reading by Starlight: Postmodern Science Fiction.* London: Routledge, 1995. Print.
Brown, Theodore. *Making Truth: Metaphor in Science.* Urbana: U of Illinois P, 2003. Print.
Buggé, Carole. *Strings.* Miller and Verma 185–254. Print.
Bush, Vannevar. *Science—The Endless Frontier.* Washington, D.C.: National Science Foundation, 1960. Print.
Callender, Craig. *Physics Meets Philosophy at the Planck Scale: Contemporary Theories in Quantum Gravity.* Cambridge: Cambridge UP, 2001. Print.
Cameron, Iain, and David O. Edge. *Scientific Images and Their Social Uses: An Introduction to the Concept of Scientism.* Boston: Butterworth, 1979. Print.
Capra, Fritjof. *The Tao of Physics: An Exploration of the Parallels between Modern Physics and Eastern Mysticism.* Berkeley: Shambhala, 1975. Print.
Castells, Manuel. *The Rise of the Network Society.* Oxford: Blackwell, 1996. Vol. 1 of *The Information Age: Economy, Society, and Culture.* 3 vols. 1996–98. Print.
Cavelos, Jeanne. *The Science of* Star Wars: *An Astrophysicist's Independent Examination of Space Travel, Aliens, Planets, and Robots As Portrayed in the* Star Wars *Films and Books.* New York: St. Martin's, 1999. Print.
Chapman, Matthew. "God or Gorilla: A Darwin Descendant at the Dover Monkey Trial." *Harper's Magazine* Feb. 2006: 54–63. Print.

Chimisso, Cristina. *Gaston Bachelard: Critic of Science and the Imagination.* London: Routledge, 2001. Print.

Chown, Marcus. "It Came from Another Dimension: First Evidence of Superstrings?" *New Scientist* Dec. 2004: 42–44. Print.

Clarke, Bruce. *Energy Forms: Allegory and Science in the Era of Classical Mechanics.* Ann Arbor: Michigan UP, 2001. Print.

Clarke, Bruce, and Linda Dalrymple Henderson, eds. *From Energy to Information: Representation in Science and Technology, Art, and Literature.* Stanford: Stanford UP, 2002. Print.

Collins, Paul. "Must-Geek TV: Is the World Ready for an Asperger's Sitcom?" *Slate.com.* Slate Group, 6 Feb. 2009. Web. 13 Feb. 2009. http://www.slate.com/id/2210635/.

Comte, Auguste. *Auguste Comte and Positivism: The Essential Writings.* Ed. Gertrud Lenzer. New York: Harper & Row, 1975. Print.

Connes, Alain, and S.K. Berberian. *Noncommutative Geometry.* New York: Academic P, 1994. Print.

Connor, Steven. *The Book of Skin.* London: Reaktion, 2004. Print.

Connor, Steven. "Topologies: Michel Serres and the Shapes of Thought." *Anglistik* 15 (2004): 105–17. Print.

Cordle, Daniel. *Postmodern Postures: Literature, Science and the Two Cultures Debate.* Aldershot, UK: Ashgate, 1999. Print.

Cornell, Drucilla. *Background Interference.* n.d. TS.

Cornell University lanl.arXiv.org e-Print Archive Mirror. Cornell U Library, n.d. Web. 1 June 2009. http://arxiv.org/.

Davies, Paul, and Julian R. Brown, eds. *Superstrings: A Theory of Everything?* Cambridge: Cambridge UP, 1988. Print.

Davis, Graham, and Hugh and Nancy Diamond. *The Rise of American Research Universities: Elites and Challengers in the Postwar Era.* Baltimore: Johns Hopkins UP, 1997. Print.

De Beaugrande, Robert. "Quantum Aspects of Perceived Reality: A New Engagement of Science and Art." *Journal of Literary Semantics* 18 (1989): 1–49. Print.

Dear, Peter, ed. *The Literary Structure of Scientific Argument.* Philadelphia: U of Penn P, 1990. Print.

Deser, S. "Infinities in Quantum Gravities." *Annalen der Physik* 9 (1999): 299–307. *ArXiv.org.* Web. 1 June 2009. gr-qc/9911073v1.

Deser, S. "Nonrenormalizability of (Last Hope) D=11 Supergravity, with a Terse Survey of Divergences in Quantum Gravities." *Fundamental Interactions: From Symmetries to Black Holes.* Brussels. March 1999. Lecture. *ArXiv.org.* Web. 1 June 2009. hep-th/9905017v1.

Deutscher, Max, ed. *Michèle Le Doeuff: Operative Philosophy and Imaginary Practice.* Amherst, NY: Humanity, 2000. Print.

Ede, Sian, ed. *Strange and Charmed: New Science and the Contemporary Visual Arts.* London: Gulbenkian Foundation, 2000. Print.

Einstein, Albert. *Relativity: The Special and the General Theory, a Popular Exposition.* Trans. Robert P. Lawson. 15th ed. London: Methuen, 1954. Print.

Einstein, A., P. Podolsky, and N. Rosen. "Can a Quantum-Mechanical Description of Physical Reality Be Considered Complete?" *Physical Review* 47 (1930): 777–80. Print.

"The Elegant Universe: A Three-Hour Miniseries with Brian Greene." Narr. Brian Greene. Dirs. Julia Cort and Joseph McMaster. *NOVA*. PBS. WGBH, Boston, 28 Oct.–4 Nov. 2003. Web. 1 June 2009.

European Organization for Nuclear Research. *Large Hadron Collider Official Web Site*. CERN, n.d. Web. 2 June 2009. http://lhc.web.cern.ch/lhc/.

Fauconnier, Gilles, and Mark Turner. *The Way We Think: Conceptual Blending and the Mind's Hidden Complexities*. New York: Basic Books, 2002. Print.

Feyerabend, Paul. *Against Method*. London: Verso, 1993. Print.

Fludernik, Monika. "Histories of Narrative Theory (II): From Structuralism to the Present." *A Companion to Narrative Theory*. Eds. James Phelan and Peter J. Rabinowitz. Oxford: Blackwell, 2005. Print.

Foucault, Michel. *The Order of Things: An Archaeology of the Human Sciences*. New York: Routledge, 2002. Print.

Frayn, Michael. *Copenhagen*. London: Methuen Drama, 1998. Print.

Freud, Sigmund. *The Uncanny*. Trans. David McLintock. London: Penguin, 2003. Print.

Frye, Northrop. *Anatomy of Criticism: Four Essays*. Princeton: Princeton UP, 1957. Print.

Gale, George, and Cassandra L. Pinnick. "Stalking Theoretical Physicists: An Ethnography Flounders: A Response to Merz and Knorr Cetina." *Social Studies of Science* 27.1 (1997): 113–23. Print.

Galilei, Galileo. "The Assayer." *Discoveries and Opinions of Galileo*. Trans. Stillman Drake. New York: Doubleday, 1957. 237–38. Print.

Galison, Peter. *How Experiments End*. Chicago: U of Chicago P, 1987. Print.

Galison, Peter. *Image and Logic: A Material Culture of Microphysics*. Chicago: U of Chicago P, 1997. Print.

Gates, Jr., S. James. *Superstring Theory: The DNA of Reality*. 4 discs. Chantilly, VA: Teaching Co., 2006. DVD.

Gefter, Amanda. "Is String Theory In Trouble?: An Interview with Leonard Susskind." *New Scientist* Dec. 2005: 48. Print.

Gieryn, Thomas F. *Cultural Boundaries of Science: Credibility on the Line*. Chicago: U of Chicago P, 1999. Print.

Ginsberg, Allen. "Sunflower Sutra." *Howl and Other Poems*. San Francisco: City Lights, 1959. 35–38. Print.

Gliozzi, F., J. Scherk, and D. Olive. "Supergravity and the Spinor Dual Model." *Physics Letters* 65B (1976): 282–86. Print.

Goddard, P., J. Goldstone, C. Rebbi, and Charles B. Thorn. "Quantum Dynamics of a Massless, Relativistic String." *Nuclear Physics* B56 (1973): 109–35. Print.

Green, Michael B. "Superstrings." *Scientific American* Sept. 1986: 44–56. Print.

Green, Michael B. "Superstrings and the Unification of Physical Forces." *Mathematical Physics 2000*. Eds. A. Fokas, T.W.B. Kibble, A. Grigouriou, and B. Zegarlinski. London: Imperial College P, 2000. 59–86. Print.

Green, Michael B., and John H. Schwarz. "Anomaly Cancellation in Supersymmetric D=10 Gauge Theory and Superstring Theory." *Physics Letters* B149 (1984): 117–22. Print.

Green, Michael B., and John H. Schwarz. "Superstring Field Theory." *Nuclear Physics* B243 (1984): 475–536. Print.

Green, Michael B., and John H. Schwarz. "Supersymmetric Dual String Theory." *Nuclear Physics* B181 (1981): 502–30. Print.

Green, Michael B., John H. Schwarz, and Edward Witten. *Superstring Theory.* 2 vols. Cambridge: Cambridge UP, 1987. Print.

Greene, Brian. *The Elegant Universe: Superstrings, Hidden Dimensions, and the Quest for the Ultimate Theory.* New York: Random House, 1999. Print.

Greene, Brian. *The Fabric of the Cosmos: Space, Time, and the Texture of Reality.* New York: Random House, 2004. Print.

Gregory, Bruce. *Inventing Reality: Physics as Language.* New York: Wiley, 1988. Print.

Gribbin, John. *The Search for Superstrings, Symmetry, and the Theory of Everything.* London: Little, Brown, 1998. Print.

Gross, D. J., J. A. Harvey, E. Martinec, and R. Rohm. "Heterotic String Theory (I): The Free Heterotic String." *Nuclear Physics* B256 (1985): 253–84. Print.

Gross, Paul, and Norman Levitt. *Higher Superstition: The Academic Left and its Quarrels with Science.* Baltimore: Johns Hopkins UP, 1994. Print.

Gross, Paul, Norman Levitt, and Martin Lewis. *The Flight from Science and Reason.* Baltimore: Johns Hopkins UP, 1997. Print.

Hakfoort, Casper. "The Historiography of Scientism: A Critical Review." *History of Science* 33.102 (1995): 375–95. Print.

Halpern, Paul. *The Great Beyond: Higher Dimensions, Parallel Universes and the Extraordinary Search for a Theory of Everything.* New York: Wiley, 2004. Print.

Hartwell, David G., and Kathryn Cramer, eds. *The Ascent of Wonder: The Evolution of Hard Science Fiction.* New York: Macmillan, 1997. Print.

Hartwell, David G., and Kathryn Cramer, eds. *Year's Best SF 11.* New York: Eos, 2006. Print.

Hawking, Stephen. *A Brief History of Time: From the Big Bang to Black Holes.* New York: Bantam, 1988. Print.

Hawking, Stephen, and Leonard Mlodinow. *A Briefer History of Time.* New York: Bantam, 2005. Print.

Hayles, N. Katherine. *Chaos Bound: Orderly Disorder in Contemporary Literature and Science.* Ithaca: Cornell UP, 1990. Print.

Hayles, N. Katherine. *The Cosmic Web: Scientific Field Models and Literary Strategies in the Twentieth Century.* Ithaca: Cornell UP, 1984. Print.

Henderson, Linda Dalrymple. *The Fourth Dimension and Non-Euclidean Geometry in Modern Art.* Princeton: Princeton UP, 1983. Print.

Herman, David. *Story Logic: Problems and Possibilities of Narrative.* Lincoln: U of Nebraska P, 2002. Print.

Hilgartner, Stephen. "The Dominant View of Popularization: Conceptual Problems, Political Uses." *Social Studies of Science* 20.3 (1990): 519–39. Print.

Hillman, Brenda. "On Song, Lyric, and Strings." *NewAmericanWriting.com.* OINK! Press: Mill Valley, CA, 2006. Web. 11 June 2009. http://newamericanwriting.com/25/hillman.htm.

Hillman, Brenda. "String Theory Sutra." Miller and Verma 168–70. Print.

Hollinger, Veronica, and Joan Gordon, eds. *Edging into the Future: Science Fiction and Contemporary Cultural Transformation.* Philadelphia: U of Penn P, 2002. Print.

Holton, Gerald. *The Scientific Imagination.* Cambridge: Cambridge UP, 1978. Print.

Hoyle, C. D., et al. "Submillimeter Test of the Gravitational Inverse-Square Law: A

Search for 'Large' Extra Dimensions." *Physics Review Letters* 86.8 (2001): 1418–21. Print.

Huggan, Graham and Helen Tiffin. *Postcolonial Ecocriticism: Literature, Animals, Environment.* Routledge, 2010. Print.

Hurtado, Christina. "Return of the (Televised) Nerds: CBS's *The Big Bang Theory* Brings Nerd Culture, With Science Chops, Back to Mainstream TV." *PopSci.com.* Popular Science, 4 Dec. 2008. Web. 13 Feb. 2009.

Jacobellis v. Ohio. 378 US 184. Supreme Court of the US. 1964. *Justia.com.* U.S. Supreme Court Center, n.d. Web. 3 June 2009.

James, Edward. *Science Fiction in the Twentieth Century.* Oxford: Oxford UP, 1994. Print.

Jarman, Heather. *Evolution: Star Trek Voyager String Theory, Book 3.* New York: Simon & Schuster, 2006. Print.

Jones, Caroline A., and Peter Galison, eds. *Picturing Art, Producing Science.* New York: Routledge, 1998. Print.

Jones, Charlotte. *Humble Boy: A Play.* London: Faber & Faber, 2003. Print.

Joyce, James. *Finnegans Wake.* London: Penguin, 1999. Print.

Kaku, Michio. *Hyperspace: A Scientific Odyssey Through Parallel Universes, Time Warps, and the 10th Dimension.* New York: Oxford UP, 1994. Print.

Kaku, Michio. *Parallel Worlds: A Journey Through Creation, Higher Dimensions, and the Future of the Cosmos.* New York: Doubleday, 2005. Print.

Kaku, Michio. *Physics of the Future: How Science Will Shape Human Destiny and Our Daily Lives by the Year 2100.* New York: Doubleday, 2011. Print.

Kaku, Michio. *Physics of the Impossible: A Scientific Exploration into the World of Phasers, Force Fields, Teleportation, and Time Travel.* New York: Doubleday, 2008. Print.

Kaku, Michio. *Visions: How Science Will Revolutionize the 21st Century.* Oxford: Oxford UP, 1999. Print.

Kaku, Michio, and Jennifer Trainer. *Beyond Einstein: The Cosmic Quest for the Theory of the Universe.* New York: Bantam, 1987. Print.

Kakushadze, Zurab, and S.-H. Henry Tye. "Brane World." *Nuclear Physics* B548 (1999): 180–204. *ArXiv.org.* Web. 13 Mar. 2008. hep-th/9809147.

Kibble, Tom. "Cosmic Strings Reborn?" COSLAB 2004, Ambleside, Cumbria, UK. 10–17 Sept. 2004. Lecture. *ArXiv.org.* Web. 2 June 2009. astro-ph/0410073v2.

Koertge, Noretta, ed. *A House Built on Sand: Exposing Postmodernist Myths About Science.* New York: Oxford UP, 1998. Print.

Koyré, Alexander. *From the Closed World to the Infinite Universe.* New York: Harper, 1958. Print.

Krauss, Lawrence. *Beyond* Star Trek: *From Alien Invasions to the End of Time.* New York: Basic Books, 1997. Print.

Krauss, Lawrence. "Discovery for the Sake of Discovery: From the New Particle Accelerator at CERN May Emerge Answers to the Most Fundamental Questions of the Universe." *Seed Magazine.com.* Seed Media Group, 31 May 2006. Web. 12 Sept. 2009.

Krauss, Lawrence. *Hiding in the Mirror: The Mysterious Allure of Extra Dimensions, from Plato to String Theory and Beyond.* New York: Viking, 2005. Print.

Krauss, Lawrence. *The Physics of* Star Wars. New York: Basic Books, 1995. Print.

Krieger, Martin H. *Doing Physics: How Physicists Take Hold of the World.* Bloomington: Indiana UP, 1992. Print.
Kuhn, Thomas. *The Structure of Scientific Revolutions.* Chicago: U of Chicago P, 1996. Print.
Lakoff, George. *Women, Fire, and Dangerous Things: What Categories Reveal About the Mind.* Chicago: U of Chicago P, 1987. Print.
Lakoff, George, and Mark Johnson. *Metaphors We Live By.* Chicago: U of Chicago P, 1980. Print.
Lakoff, George, and Mark Johnson. *Philosophy in the Flesh: The Embodied Mind and Its Challenge to Western Thought.* New York: Basic Books, 1999. Print.
Lakoff, George, and Rafael Núñez. *Where Mathematics Comes From: How the Embodied Mind Brings Mathematics Into Being.* New York: Basic Books, 2001. Print.
Lakoff, George, and Mark Turner. *More Than Cool Reason: A Field Guide to Poetic Metaphor.* Chicago, IL: U of Chicago P, 1989. Print.
Lang, Jeffrey. *Cohesion: Star Trek Voyager String Theory, Book 1.* New York: Simon & Schuster, 2005. Print.
Latour, Bruno. "A Relativistic Account of Einstein's Relativity." *Social Studies of Science* 18.1 (1988): 3–44. Print.
Laughlin, Robert B. *A Different Universe: Reinventing Physics from the Bottom Up.* New York: Basic Books, 2005. Print.
Le Doeuff, Michèle. *The Philosophical Imaginary.* Trans. Colin Gordon. London: Continuum, 2002. Print.
Leane, Elizabeth. *Reading Popular Physics: Disciplinary Skirmishes and Textual Strategies.* Aldershot, UK: Ashgate, 2007. Print.
Lenoir, Timothy, ed. *Inscribing Science: Scientific Texts and the Materialities of Communication.* Stanford: Stanford UP, 1998. Print.
Lessl, Thomas M. "The Priestly Voice." *Quarterly Journal of Speech* 75 (1989): 183–97. Print.
Lessl, Thomas M. "Science and the Sacred Cosmos: The Ideological Rhetoric of Carl Sagan." *Quarterly Journal of Speech* 71 (1985): 175–87. Print.
Lovelace, C. "Pomeron Form Factors and Dual Regge Cuts." *Physics Letters* 34B (1971): 500–506. Print.
Luckhurst, Roger. *Science Fiction.* London: Polity, 2005. Print.
Lucretius. *De Rerum Natura.* Trans. H.A.J. Munro. Cambridge: Liber I, 1886. Print.
Luhmann, Niklas. "The Cognitive Program of Constructivism and the Reality That Remains Unknown." *Theories of Distinction: Redescribing the Descriptions of Modernity.* Trans. Joseph O'Neil, Elliott Schreiber, Kerstin Behnke, and William Whobrey. Stanford: Stanford UP, 2002. 128–52. Print.
Malcohn, Elissa. "Arachne." Miller and Verma 168–70. Print.
Maldacena, Juan. "The Large N Limit of Superconformal Field Theories and Supergravity." *Advances in Theoretical and Mathematical Physics* 2 (1998): 231–52. *ArXiv.org.* Web. 2 June 2009. hep-th/9711200v3.
Marolf, Don. "Resource Letter NSST-1: The Nature and Status of String Theory." *American Journal of Physics* 72 (2004): 730–41. *ArXiv.org.* Web. 2 June 2009. hep-th/0311044v5.
McBurney, Simon. *A Disappearing Number.* London: Oberon, 2008. Print.

Mellor, Felicity. "Between Fact and Fiction: Demarcating Science from Non-Science in Popular Physics Books." *Social Studies of Science* 33.4 (2003): 509–38. Print.

Merz, Martina, and Karin Knorr Cetina. "Deconstruction in a 'Thinking' Science: Theoretical Physicists at Work." *Social Studies of Science* 27.1 (1997): 73–111. Print.

Midgley, Mary. *Science and Poetry.* London: Routledge, 2001. Print.

Miller, Sean. "Imagining Braneworlds in String Theory Technical Discourse." *Discourse* 29.1 (2007): 77–100. Print.

Miller, Sean, and Shveta Verma. *Riffing on Strings: Creative Writing Inspired by String Theory.* New York: Scriblerus, 2008. Print.

Mohaupt, Thomas. "Introduction to String Theory." *Lecture Notes in Physics* 631 (2003): 173–251. *ArXiv.org.* Web. 3 June 2009. hep-th/0207249v1.

Nadeau, Robert. *Readings from the New Book on Nature: Physics and Metaphysics in the Modern Novel.* Amherst: U of Mass P, 1981. Print.

Nambu, Yoichiro. "Quark Model and the Factorization of the Veneziano Amplitude." *Symmetries and Quark Models.* Ed. Ramesh Chand. New York: Gordon & Breach, 1970. 269–78. Print.

Nambu, Yoichiro, and Paul Frampton. "Asymptotic Behavior of Partial Widths in the Veneziano Model of Scattering Amplitudes." *Quanta: Essays in Theoretical Physics Dedicated to Gregor Wentzel.* Eds. P.G.O. Freud, C. J. Goebel, and Y. Nambu. Chicago: U of Chicago P, 1970. 403–14. Print.

Newton, Isaac. *Sir Isaac Newton's Mathematical Principles of Natural Philosophy and His System of the World.* Trans. Andrew Motte and Florian Cajori. Berkeley: U of California P, 1962. Print.

Orr, Mary. *Intertextuality: Debates and Contexts.* Cambridge: Polity, 2003. Print.

Ortony, Andrew. *Metaphor and Thought.* 2nd ed. Cambridge: Cambridge UP, 1993. Print.

Osserman, Robert. *Poetry of the Universe: A Mathematical Exploration of the Cosmos.* London: Anchor, 1995. Print.

The Oxford English Dictionary. 2nd ed. Oxford: Clarendon, 1989. Print.

Peat, F. David. *Superstrings and the Search for the Theory of Everything.* Chicago: Contemporary Books, 1988. Print.

Penrose, Roger. *The Road to Reality: A Complete Guide to the Laws of the Universe.* New York: Knopf, 2005. Print.

Petcher, Donald N. Personal interview. 9 Dec. 2010.

Pickering, Andrew. *Constructing Quarks: A Sociological History of Particle Physics.* Chicago: U of Chicago P, 1984. Print.

Pickering, Andrew. *The Mangle of Practice: Time, Agency, and Science.* Chicago: U of Chicago P, 1995. Print.

Pickover, Clifford A. *Surfing Through Hyperspace: Understanding Higher Universes in Six Easy Lessons.* New York: Oxford UP, 1999. Print.

Polchinski, Joseph. "Dirichlet-Branes and Ramond-Ramond Charges." *Physical Review Letters* 75 (1995): 4724–7. *ArXiv.org.* Web. 2 June 2009. hep-th/9510017v3.

Polchinski, Joseph. *String Theory.* 2 vols. Cambridge: Cambridge UP, 1999. Print.

Polyakov, A. M. "Quantum Geometry of Bosonic Strings." *Physics Letters* 103B (1981): 207–10. Print.

Polyakov, A. M. "Quantum Geometry of Fermionic Strings." *Physics Letters* 103B (1981): 211–13. Print.

Popper, Karl. *The Logic of Scientific Discovery.* London: Hutchinson, 1959. Print.

Porter, Noah, ed. *Webster's International Dictionary of the English Language.* Springfield, MA: G & C Merriam, 1900. Print.

Porush, David. "Cybernetic Fiction and Postmodern Science." *New Literary History* 20 (1989): 373–96. Print.

Porush, David. "Voyage to Eudoxia: The Emergence of a Post-Rational Epistemology in Literature and Science." *Sub-Stance* 71.2 (1993): 38–49. Print.

Quine, Willard Van Orman. *From a Logical Point of View.* 2nd rev. ed. Cambridge: Harvard UP, 1980. Print.

Randall, Lisa. *Warped Passages: Unraveling the Mysteries of the Universe's Hidden Dimensions.* New York: Ecco, 2005. Print.

Randall, Lisa, and Raman Sundrum. "An Alternative to Compactification." *Physical Review Letters* 83 (1999): 4690–3. *ArXiv.org.* Web. 12 February 2008. hep-th/9906064.

Randall, Lisa, and Raman Sundrum. "A Large Mass Hierarchy from a Small Extra Dimension." *Physical Review Letters* 83 (1999): 3370–373. *ArXiv.org.* Web. 3 June 2009. hep-ph/9905221v1.

Randall, Lisa, and Raman Sundrum. "Out Of This World Supersymmetry Breaking." *Nuclear Physics* B557 (1999): 79–118. *ArXiv.org.* Web. 3 June 2009. hep-th/9810155v2.

Reichenbach, Hans. *The Rise of Scientific Philosophy.* Berkeley: U of California P, 1951. Print.

Reingold, Jacquelyn. *String Fever.* New York: Dramatists Play Service, 2003. Print.

Ricoeur, Paul. *The Rule of Metaphor: Multi-disciplinary Studies of the Creation of Meaning in Language.* Trans. Robert Czerny. Toronto: U of Toronto P, 1977. Print.

Roberts, Adam. *The History of Science Fiction.* New York: Palgrave, 2006. Print.

Roberts, Adam. "S-Bomb." Miller and Verma 31–46. Print.

Roberts, Adam. *Science Fiction.* 2nd ed. London: Routledge, 2005. Print.

Rothblatt, Sheldon, and Björn Wittrock, eds. *The European and American University Since 1800: Historical and Sociological Essays.* Cambridge: Cambridge UP, 1993. Print.

Rucker, Rudy. *The Fourth Dimension.* Boston: Houghton Mifflin, 1984. Print.

Rucker, Rudy. *Geometry, Relativity, and the Fourth Dimension.* New York: Dover, 1977. Print.

Rucker, Rudy. "Guadalupe and Hieronymus Bosch." Hartwell and Cramer 89–105. Print.

Russell, Bertrand. *The ABC of Relativity.* 5th rev. ed. London: Routledge, 1997. Print.

Sacks, Sheldon, ed. *On Metaphor.* Chicago: U of Chicago P, 1979. Print.

Sayer, Karen, and John Moore, eds. *Science Fiction, Critical Frontiers.* Basingstoke, UK: Macmillan, 2000. Print.

Scherk, Joel, and John H. Schwarz. "Dual Models for Nonhadrons." *Nuclear Physics* B81 (1974): 118–44. Print.

Schwarz, John H. "Introduction to Superstring Theory." NATO Advanced Study

Institute on Techniques and Concepts of High Energy Physics, St. Croix. 9 Aug. 2000. Lecture. *ArXiv.org*. Web. 2 June 2009. hep-ex/0008017v1.

Schwarz, John H. "String Theory." *Current Science* 81.12 (2001): 1547–53. Print.

Schwarz, John H. "String Theory: The Early Years." Symposium for L. Susskind's 60th Birthday. Stanford U, Palo Alto, CA. 20–21 May 2000. Lecture. *ArXiv.org*. Web. 2 June 2009. hep-th/0007118v3.

Serres, Michel. *The Birth of Physics*. Trans. Jack Hawkes. Manchester, UK: Clinamen, 2000. Print.

Serres, Michel. *Genesis*. Trans. Genevieve James and James Nielson. Ann Arbor: U of Michigan P, 1997. Print.

Serres, Michel. *The Troubadour of Knowledge*. Trans. Sheila Faria Glaser. Ann Arbor: U of Michigan P, 1997. Print.

Serres, Michel, and Bruno Latour. *Conversations on Science, Culture, and Time*. Trans. Roxanne Lapidus. Ann Arbor: U of Michigan P, 1995. Print.

Shapin, Steven, and Simon Schaffer. *Leviathan and the Air-Pump: Hobbes, Boyle, and the Experimental Life*. Princeton: Princeton UP, 1985. Print.

Shepherd-Barr, Kirsten. *Science on Stage: From* Doctor Faustus *to* Copenhagen. Princeton: Princeton UP, 2006. Print.

Shermer, Michael. "The Shamans of Scientism." *Scientific American* June 2002: 20. Print.

Silverman, Kaja. *The Subject of Semiotics:* Oxford: Oxford UP, 1983. Print.

Smelik, Anneke, ed. *The Scientific Imaginary in Visual Culture*. Göttingen: Vandenhoeck and Ruprecht, 2010. Print.

Smolin, Lee. "How Far Are We from the Quantum Theory of Gravity?" 11 Apr. 2003. *ArXiv.org*. Web. 2 June 2009. hep-th/0303185v2.

Smolin, Lee. *Three Roads to Quantum Gravity*. New York: Basic Books, 2001. Print.

Smolin, Lee. *The Trouble with Physics: The Rise of String Theory and the Fall of a Science, and What Comes Next*. New York: Houghton Mifflin, 2006. Print.

Snow, C. P. *The Two Cultures*. Cambridge: Cambridge UP, 1959. Print.

Sokal, Alan. "Transgressing the Boundaries: Towards a Transformative Hermeneutics of Quantum Gravity." *Social Text* 46–47 (1996): 217–52. Web. 10 Sept 2009. http://www.physics.nyu.edu/faculty/sokal/transgress_v2/transgress_v2_singlefile.html.

Sokal, Alan, and Jean Bricmont. *Fashionable Nonsense: Postmodern Intellectuals' Abuse of Science*. New York: Picador, 1998. Print.

Speier, Susanna. *Calabi-Yau*. 2001. TS.

Spenser, Edmund. *The Faerie Queene*. New York: Penguin, 1978. Print.

Stanford Linear Accelerator Center. *SPIRES High Energy Physics Database*. SLAC, n.d. Web. 15 Oct. 2009.

Steinhardt, Paul, and Neil Turok. *Endless Universe: Beyond the Big Bang*. New York: Doubleday, 2007. Print.

Stewart, Ian. *Nature's Numbers: The Unreal Reality of Mathematics*. New York: Basic Books, 1995. Print.

Strauss, Gary. "There's a Science to CBS' *Big Bang Theory*." *USAToday.com*. USA Today, 4 Nov. 2007. Web. 13 Feb. 2009.

Strehle, Susan. *Fiction in the Quantum Universe*. Chapel Hill: U of North Carolina P, 1992. Print.

Strehle, Susan. "To the Beat of a Different Conundrum: Postmodern Science and Literature." *Postmodern Times: A Critical Guide to the Contemporary*. Ed. Thomas Carmichael, Alison Lee, and Charles Jencks. DeKalb: Northern Illinois UP, 2000. 209–28. Print.

Strominger, Andrew, and Cumrun Vafa. "Microscopic Origin of the Bekenstein-Hawking Entropy." *Physics Letters* B379 (1996): 99–104. Print.

Susskind, Leonard. *The Cosmic Landscape: String Theory and the Illusion of Intelligent Design*. New York: Little, Brown, 2005. Print.

Susskind, Leonard. "Dual Symmetric Theory of Hadrons—I." *Nuovo Cimento* A69 (1970): 457–96. Print.

Susskind, Leonard. "Harmonic Oscillator Analogy for the Veneziano Amplitude." *Physical Review Letters* 23.545 (1969): 545–47. Print.

Susskind, Leonard. "Twenty Years of Debate with Stephen." Stephen Hawking 60th Birthday Celebration and Conference. Centre for Mathematical Sciences, Cambridge, UK. 7–11 Jan. 2002. Lecture. *ArXiv.org*. Web. 2 June 2009. hep-th/0204027v1.

Suvin, Darko. "On the Poetics of the Science Fiction Genre." *College English* 34.3 (1972): 372–82. Print.

Tabbi, Joseph. *Cognitive Fictions*. Minneapolis: U of Minnesota P, 2002. Print.

Terranova, Elaine. "String Theory." Miller and Verma 156–57. Print.

Thurtle, Phillip. *The Emergence of Genetic Rationality: Space, Time, and Information in American Biology 1870–1920*. Seattle: U of Washington P, 2008. Print.

Thurtle, Phillip, and Robert Mitchell, eds. *Data Made Flesh: Embodying Information*. London: Routledge, 2003. Print.

Traweek, Sharon. *Beamtimes and Lifetimes: The World of High Energy Physicists*. Cambridge: Harvard UP, 1988. Print.

Turney, John. "Accounting For Explanation in Popular Science Texts—An Analysis of Popularized Accounts of Superstring Theory." *Public Understanding of Science* 13 (2004): 331–46. Print.

U.S. Dept. of Energy. *Superconducting Super Collider Web Site*. Stanford Linear Accelerator Center, Palo Alto, CA, 1995. Web. 2 June 2009. http://www.hep.net/ssc/.

Vafa, Cumrun. "Evidence for F-theory." *Nuclear Physics* B469 (1996): 403–18. *ArXiv.org*. Web. 2 June 2009. hep-th/9602022v1.

Veneziano, G. "Construction of a Crossing-Symmetric, Regge-Behaved Amplitude for Linearly Rising Trajectories." *Nuovo Cimento* A57 (1968): 190–97. Print.

Vilenkin, A. *Cosmic Strings and Other Topological Defects*. Cambridge: Cambridge UP, 2000. Print.

Wald, Priscilla. *Contagious: Cultures, Carriers, and the Outbreak Narrative*. Durham: Duke UP, 2008. Print.

Webb, Stephen. *Out of This World: Colliding Universes, Branes, Strings, and Other Wild Ideas of Modern Physics*. New York: Springer, 2004. Print.

Weinberg, Steven. *Dreams of a Final Theory: The Scientist's Search for the Ultimate Laws of Nature*. New York: Pantheon, 1992. Print.

Wess, J., and B. Zumino. "Supergauge Transformations in Four Dimensions." *Nuclear Physics* B70 (1974): 39–50. Print.

White, Eric. "Contemporary Cosmology and Narrative Theory." *Literature and*

Science: Theory and Practice. Ed. Stuart Peterfreund. Boston: Northeastern UP, 1990. 91–112. Print.

Witten, Edward. "Anti de Sitter Space and Holography." *Advances in Theoretical and Mathematical Physics* 2 (1998): 253–91. *ArXiv.org.* Web. 2 June 2009. hep-th/9802150v2.

Witten, Edward. "Duality, Spacetime, and Quantum Mechanics." Kavli Institute for Theoretical Physics. UC Santa Barbara, CA. Jan. 1998. Lecture. Web. 2 June 2009. http://online.itp.ucsb.edu/online/plecture/witten/.

Witten, Edward. "Perturbative Gauge Theory as a String Theory in Twistor Space." *Communications in Mathematical Physics* 252 (2004): 189–258. *ArXiv.org.* Web. 2 June 2009. hep-th/0312171v2.

Witten, Edward. "Reflections on the Fate of Spacetime." *Physics Today* Apr. 1996: 24–30. Print.

Witten, Edward. "String Theory Dynamics in Various Dimensions." *Nuclear Physics* B443 (1995): 85–126. *ArXiv.org.* Web. 2 June 2009. hep-th/9503124v2.

Witten, Edward. "The Universe on a String." *Astronomy* June 2002: 42–47. Print.

Woit, Peter. "Is String Theory Even Wrong?" *American Scientist* Mar.–Apr. 2002: 110. Print.

Woit, Peter. *Not Even Wrong: The Failure of String Theory and the Continuing Challenge to Unify the Laws of Physics.* London: Jonathan Cape, 2006. Print.

Woolgar, Steve, and Michael Lynch, eds. *Representation in Scientific Practice.* Cambridge: MIT Press, 1990. Print.

Wutz, Michael. *Enduring Words: Literary Narrative in Changing Media Ecology.* Tuscaloosa: U of Alabama P, 2009. Print.

Wutz, Michael, and Joseph Tabbi, eds. *Reading Matters: Narrative in the New Media Ecology.* Ithaca: Cornell UP, 1997. Print.

Zee, A. *Quantum Field Theory in a Nutshell.* Princeton: Princeton UP, 2003. Print.

Zunshine, Lisa, ed. *Introduction to Cognitive Cultural Studies.* Baltimore: Johns Hopkins UP, 2010. Print.

Zwiebach, Barton. *A First Course in String Theory.* Cambridge: Cambridge UP, 2004. Print.

Index